COMPUTER ALGEBRA 2006
Latest Advances in Symbolic Algorithms

COMPUTER ALGEBRA 2006

Latest Advances in Symbolic Algorithms

Proceedings of the Waterloo Workshop in
Computer Algebra 2006

Ontario, Canada   10 – 12 April 2006

Editors

Ilias Kotsireas
Eugene Zima

Wilfrid Laurier University, Canada

 **World Scientific**

NEW JERSEY · LONDON · SINGAPORE · BEIJING · SHANGHAI · HONG KONG · TAIPEI · CHENNAI

Published by

World Scientific Publishing Co. Pte. Ltd.

5 Toh Tuck Link, Singapore 596224

USA office: 27 Warren Street, Suite 401-402, Hackensack, NJ 07601

UK office: 57 Shelton Street, Covent Garden, London WC2H 9HE

British Library Cataloguing-in-Publication Data
A catalogue record for this book is available from the British Library.

COMPUTER ALGEBRA 2006
Latest Advances in Symbolic Algorithms

ISBN-13 978-981-270-200-5
ISBN-10 981-270-200-8

Printed in Singapore.

Sergei Abramov

PREFACE

The Waterloo Workshop on Computer Algebra (WWCA-2006) was held on April 10–12, 2006 at Wilfrid Laurier University (Waterloo, Ontario, Canada) hosted by CARGO (http://www.cargo.wlu.ca). The workshop provided a forum for researchers and practitioners to discuss recent advances in the area of Computer Algebra. WWCA-2006 was dedicated to the 60th birthday of Sergei Abramov (Computer Center of the Russian Academy of Sciences, Moscow, Russia) whose influential contributions to symbolic methods are highly acknowledged by the research community and adopted by the leading Computer Algebra systems. The workshop attracted world-renowned experts from both the academia and the software industry. Presentations on original research topics or surveys of the state of the art advances in particular areas of Computer Algebra were made by

- Sergei Abramov, CCRAS, Russia
- Moulay Barkatou, University of Limoges, France
- Jacques Carette, McMaster University, Canada
- Robert Corless, University of Western Ontario, Canada
- Jürgen Gerhard, Maplesoft, Canada
- Oleg Golubitsky, Queens University, Canada
- Gaston Gonnet, ETH Zurich, Switzerland
- Kevin Hare, University of Waterloo, Canada
- Ilias Kotsireas, Wilfrid Laurier University, Canada
- George Labahn, University of Waterloo, Canada
- Ziming Li, Academy of Mathematics and System Sciences, China
- Luc Rebillard, University of Waterloo, Canada
- Bruno Salvy, INRIA Rocquencourt, France
- Éric Schost, University of Western Ontario, Canada
- Arne Storjohann, University of Waterloo, Canada
- Serguei Tsarev, Krasnoyarsk State Pedagogical University, Russia
- Mark van Hoeij, Florida State University, USA
- Thomas Wolf, Brock University, Canada
- Doron Zeilberger, Rutgers University, USA
- Eugene Zima, Wilfrid Laurier University, Canada

Presentations abstracts are published by the ACM Communications in Computer Algebra, Volume 40, Number 2, June 2006, Issue 156, pp. 52-59.

Success of the workshop was also due to the support of the Academic Development Fund, Office of the Vice-President Academic, Research Office, and Department of Physics and Computer Science of Wilfrid Laurier University.

This book presents a collection of formally refereed selected papers submitted after workshop. Topics discussed in this book are the latest advances in algorithms of symbolic summation, factorization, symbolic-numeric linear algebra and linear functional equations, i.e. topics of symbolic computations that were extensively advanced due to Sergei's influencial works.

This book wouldn't have been possible without the contributions and hard work of the anonymous referees, who supplied detailed referee reports and helped authors improve their papers significantly.

Eugene Zima,   Ilias Kotsireas
Wilfrid Laurier University,
75 University Avenue West,
Waterloo, Ontario, Canada N2L 3C5   April 2007

CONTENTS

HYPERGEOMETRIC SUMMATION REVISITED

S. A. ABRAMOV*

Dorodnicyn Computing Centre, Russian Academy of Sciences,
Vavilova 40, Moscow GSP-1, 119991, Russia
E-mail: sabramov@ccas.ru

M. PETKOVŠEK[†]

Faculty of Mathematics and Physics, University of Ljubljana,
Jadranska 19, SI-1000 Ljubljana, Slovenia
E-mail: marko.petkovsek@fmf.uni-lj.si

We consider hypergeometric sequences, i.e., the sequences which satisfy linear first-order homogeneous recurrence equations with relatively prime polynomial coefficients. Some results related to necessary and sufficient conditions are discussed for validity of discrete Newton-Leibniz formula $\sum_{k=v}^{w} t(k) = u(w + 1) - u(v)$ when $u(k) = R(k)t(k)$ and $R(k)$ is a rational solution of Gosper's equation.

Keywords: Symbolic summation; Hypergeometric sequences; Discrete Newton-Leibniz formula.

1. Introduction

Let K be a field of characteristic zero ($K = \mathbb{C}$ in all examples). If $t(k) \in K(k)$ then the *telescoping equation*

$$u(k + 1) - u(k) = t(k) \tag{1}$$

may or may not have a rational solution $u(k)$, depending on the type of $t(k)$. Here the telescoping equation is considered as an equality in the rational-function field, regardless of the possible integer poles that $u(k)$ and/or $t(k)$ might have.

An algorithm for finding rational $u(k)$ was proposed in 1971 (see Ref. 1). It follows from that algorithm that if $t(k)$ has no integer poles, then a

---

*Partially supported by RFBR under grant 07-01-00482-a.
[†]Partially supported by MVZT RS under grant P1-0294.

rational $u(k)$ satisfying (1), if it exists, has no integer poles either, and the *discrete Newton-Leibniz formula*

$$\sum_{k=v}^{w} t(k) = u(w+1) - u(v) \tag{2}$$

is valid for any integer bounds $v \leq w$. Working with polynomial and rational functions we will write $f(k) \perp g(k)$ for $f(k), g(k) \in K[k]$ to indicate that $f(k)$ and $g(k)$ are coprime; if $R(k) \in K(k)$, then $\text{den}(R(k))$ is the monic polynomial from $K[k]$ such that $R(k) = \frac{f(k)}{\text{den}(R(k))}$ for some $f(k) \in K[k], f(k) \perp \text{den}(R(k))$.

The problem of solving equation (1) can be considered for sequences. If $t(k)$ is a sequence, we use the symbol E for the shift operator w.r. to k, so that $Et(k) = t(k+1)$. In the rest of the paper we assume that the sequences under consideration are defined on an infinite interval I of integers and either $I = \mathbb{Z}$, or

$$I = \mathbb{Z}_{\geq l} = \{k \in \mathbb{Z} \mid k \geq l\}, \quad l \in \mathbb{Z}.$$

If a sequence $t(k)$ defined on I is given, and a sequence $u(k)$, which is also defined on I and satisfies (1) for all $k \in I$, is found (any such sequence is a *primitive* of $t(k)$), then we can use formula (2) for any $v \leq w$ with $v, w \in I$.

Gosper's algorithm,[6] which we denote hereafter by $\mathcal{G}\mathcal{A}$, discovered in 1978, focuses on the case where a given $t(k)$ and an unknown $u(k)$ are hypergeometric sequences.

Definition 1.1. A sequence $y(k)$ defined on an infinite interval I is hypergeometric if it satisfies the equation $Ly(k) = 0$ for all $k \in I$, with

$$L = a_1(k)E + a_0(k) \in K[k, E], \quad a_1(k) \perp a_0(k). \tag{3}$$

$\mathcal{G}\mathcal{A}$ starts by constructing the operator L for a given concrete hypergeometric sequence $t(k)$, and this step is not formalized. On the next steps $\mathcal{G}\mathcal{A}$ works with L only, while the sequence $t(k)$ itself is ignored (more precisely, in the case of $L = a_1(k)E + a_0(k)$, $\mathcal{G}\mathcal{A}$ works with the *certificate* of $t(k)$, i.e., with the rational function $-\frac{a_0(k)}{a_1(k)}$, but this is not essential). The algorithm tries to construct a rational function $R(k)$, which is a solution in $K(k)$ of *Gosper's equation*

$$a_0(k)R(k+1) + a_1(k)R(k) = -a_1(k) \tag{4}$$

(such $R(k)$, when it exists, can also be found by general algorithms from Refs. 2,3). If such $R(k)$ exists then

$$R(k+1)t(k+1) - R(k)t(k) = t(k)$$

is valid for *almost* all integers k. The fact is that even when $t(k)$ is defined everywhere on I, it can happen that $R(k)$ has some poles belonging to I, and $u(k) = R(k)t(k)$ cannot be defined in such a way as to make (1) valid for all integers from I. One can encounter the situation where formula (2) is not valid even when all of

$$t(v), t(v+1), \ldots, t(w), \quad u(v), u(w+1)$$

are well-defined. The reason is that (1) may fail to hold at certain points k of the summation interval. However, sometimes it is possible to define the values of $u(k) = R(k)t(k)$ appropriately for all integers k, even though $R(k)$ has some integer poles. In such well-behaved cases (2) can be used to compute $\sum_{k=v}^{w} t(k)$ for any $v \le w$, $v, w \in I$.

Example 1.1.

Gosper's equation, corresponding to $L = kE - (k+1)^2$, has a solution $R = \frac{1}{k}$. The sequences

$$t_1(k) = \begin{cases} 0, & \text{if } k < 0, \\ k \cdot k!, & \text{if } k \ge 0 \end{cases}$$

and

$$t_2(k) = \begin{cases} \frac{(-1)^k k}{(-k-1)!}, & \text{if } k < 0, \\ 0, & \text{if } k \ge 0 \end{cases}$$

both satisfy $Ly = 0$ on $I = \mathbb{Z}$.

Generally speaking, (2) is not applicable to $t_1(k)$, but is applicable to $t_2(k)$. We can illustrate this as follows. Applying (2) to $t_1(k)$ with $v = -1, w = 1$, we have

$$t_1(-1) + t_1(0) + t_1(1) = \frac{1}{k} t_1(k) \,|_{k=2} - \frac{1}{k} t_1(k) \,|_{k=-1} = \frac{1}{2} \cdot 4 - 0 = 2$$

which is wrong, because $t_1(-1) + t_1(0) + t_1(1) = 0 + 0 + 1 = 1$. Applying (2) to t_2 with the same v, w, we have

$$t_2(-1) + t_2(0) + t_2(1) = \frac{1}{k} t_2(k) \,|_{k=2} - \frac{1}{k} t_2(k) \,|_{k=-1} = 0 - (-1) = 1$$

which is correct, because $t_2(-1) + t_2(0) + t_2(1) = 1 + 0 + 0 = 1$.

In this paper we discuss some results related to necessary and sufficient conditions for validity of formula (2) when $u(k) = R(k)t(k)$, and $R(k)$ is a rational solution of corresponding Gosper's equation. If such $R(k)$ exists, then we describe the linear space of all hypergeometric sequences $t(k)$ that are defined on I and such that formula (2) is valid for $u = Rt$ and any integer bounds $v \leq w$ such that $v, w \in I$. The dimension of this space is always positive (it can be even bigger than 1). We will denote

- by $\mathcal{H}_I$ the set of all hypergeometric sequences defined on I;
- by $\mathcal{L}$ the set of all operators of type (3);
- by $V_I(L)$, where $L \in \mathcal{L}$, the K-linear space of all sequences $t(k)$ defined on I for which $Lt(k) = 0$ for all $k \in I$;
- by $W_I(R(k), L)$, where $L \in \mathcal{L}$ and $R(k) \in K(k)$ is a solution of the corresponding Gosper's equation, the K-linear space of all $t(k) \in V_I(L)$ such that (2) with $u(k) = R(k)t(k)$ is valid for all $v \leq w$ with $v, w \in I$.

The paper is a summary of the results that have been published in Refs. 4,5. In addition we consider the case where Gosper's equation has non-unique rational solution (Section 3.2). In Section 2 we consider individual hypergeometric sequences while in Section 3 we concentrate on spaces of the type $W_I(R(k), L)$.

2. Validity conditions of the discrete Newton-Leibniz formula

2.1. *A criterion*

Theorem 2.1.[4,5] *Let* $L \in \mathcal{L}$, $t(k) \in V_I(L)$, *and let Gosper's equation corresponding to* L *have a solution* $R(k) \in K(k)$, *with* $\mathrm{den}(R) = g(k)$. *Then* $t(k) \in W_I(R(k), L)$ *iff there exists a* $\bar{t}(k) \in \mathcal{H}_I$ *such that* $t(k) = g(k)\bar{t}(k)$ *for all* $k \in I$.

Example 2.1. Consider again the sequences $t_1(k), t_2(k)$ on $I = \mathbb{Z}$ from Example 1.1. We have $t_2(k) = k\bar{t}_2(k)$, where

$$\bar{t}_2(k) = \begin{cases} \frac{(-1)^k}{(-k-1)!}, & \text{if } k < 0, \\ 0, & \text{if } k \geq 0 \end{cases}$$

is a hypergeometric sequence defined everywhere:

$$E\bar{t}_2(k) - (k+1)\bar{t}_2(k) = 0.$$

On the other hand, if $t_1(k) = k\bar{t}_1(k)$ for some sequence $\bar{t}_1(k)$, then

$$\bar{t}_1(k) = \begin{cases} 0, & \text{if } k < 0, \\ \zeta, & \text{if } k = 0, \\ k!, & \text{if } k > 0 \end{cases}$$

where $\zeta \in \mathbb{C}$. Notice that the sequence $\bar{t}_1(k)$ is not hypergeometric on $\mathbb{Z}$, for any $\zeta \in \mathbb{C}$.

2.2. *Summation of proper hypergeometric sequences*

Definition 2.1. Following conventional notation, the *rising factorial power* $(\alpha)_k$ and its *reciprocal* $1/(\beta)_k$ are defined for $\alpha, \beta \in K$ and $k \in \mathbb{Z}$ by

$$(\alpha)_k = \begin{cases} \displaystyle\prod_{m=0}^{k-1}(\alpha + m), & k \geq 0; \\ \displaystyle\prod_{m=1}^{|k|}\frac{1}{\alpha - m}, & k < 0, \ \alpha \neq 1, 2, \ldots, |k|; \\ \text{undefined}, & \text{otherwise}; \end{cases}$$

$$\frac{1}{(\beta)_k} = \begin{cases} \displaystyle\prod_{m=0}^{k-1}\frac{1}{\beta + m}, & k \geq 0, \ \beta \neq 0, -1, \ldots, 1-k; \\ \displaystyle\prod_{m=1}^{|k|}(\beta - m), & k < 0; \\ \text{undefined}, & \text{otherwise}. \end{cases}$$

Note that if $(\alpha)_k$ resp. $1/(\beta)_k$ is defined for some $k \in \mathbb{Z}$, then $(\alpha)_{k+1}$ resp. $1/(\beta)_{k-1}$ is defined for that k as well. Thus $(\alpha)_k$ and $1/(\beta)_k$ are hypergeometric sequences which satisfy

$$(\alpha)_{k+1} = (\alpha + k)(\alpha)_k, \quad (\beta + k)/(\beta)_{k+1} = 1/(\beta)_k \tag{5}$$

whenever $(\alpha)_k$ and $1/(\beta)_{k+1}$ are defined.

Example 2.2. Let $t(k) = (k-2)(-1/2)_k/(4k!)$. This hypergeometric sequence is defined for all $k \in \mathbb{Z}$ (note that $t(k) = 0$ for $k < 0$) and satisfies $Lt(k) = 0$ for all $n \in \mathbb{Z}$ where $L = a_1(k)E + a_0(k)$ with $a_0(k) = -(k-1)(2k-1)$ and $a_1(k) = 2(k-2)(k+1)$. Gosper's equation, corresponding to L, has a rational solution

$$R(k) = \frac{2k(k+1)}{k-2}. \tag{6}$$

Equation (1) indeed fails at $k = 1$ and $k = 2$ because $u(k) = R(k)t(k)$ is undefined at $k = 2$. But if we cancel the factor $k - 2$ and replace $u(k)$ by the sequence

$$\bar{u}(k) = k(k+1)\frac{(-1/2)_k}{2k!},$$

then equation

$$\bar{u}(k+1) - \bar{u}(k) = t(k) \tag{7}$$

holds for all $k \in \mathbb{Z}$, and

$$\sum_{k=v}^{w} t(k) = \bar{u}(w+1) - \bar{u}(v). \tag{8}$$

The sequence $t(k)$ from Example 2.2 is an instance of a *proper hypergeometric* sequence which we are going to define now. As it turns out, there are no restrictions on the validity of the discrete Newton-Leibniz formula for proper sequences (Theorem 2.2).

Definition 2.2. A hypergeometric sequence $t(k)$ defined on an infinite interval I of integers is *proper* if there are

- a constant $z \in K$,
- a polynomial $p(k) \in K[k]$,
- nonnegative integers q, r,
- constants $\alpha_1, \ldots, \alpha_q, \beta_1, \ldots, \beta_r \in K$

such that

$$t(k) = p(k) z^k \frac{\prod_{i=1}^{q} (\alpha_i)_k}{\prod_{j=1}^{r} (\beta_j)_k} \tag{9}$$

for all $k \in I$.

Theorem 2.2.[4] *Let $t(k)$ be a proper hypergeometric sequence defined on I and given by (9). Denote $a(k) = z \prod_{i=1}^{q} (k + \alpha_i)$ and $b(k) = \prod_{j=1}^{r} (k + \beta_j)$. If a polynomial $y(k) \in K[k]$ satisfies*

$$a(k)y(k+1) - b(k-1)y(k) = p(k) \tag{10}$$

and if

$$\bar{u}(k) = y(k) z^k \frac{\prod_{i=1}^{q} (\alpha_i)_k}{\prod_{j=1}^{r} (\beta_j)_{k-1}}$$

for all $k \in I$, then equation (7) holds for all $k \in I$, and the discrete Newton-Leibniz formula (8) is valid for all $v \leq w$, when $v, w \in I$.

Notice that (10) has a solution in $K[k]$ iff Gosper's equation, corresponding to the operator from $\mathcal{L}$, annihilating $t(k)$, has a solution in $K(k)$.

Example 2.3. The hypergeometric sequence

$$t(k) = \frac{\binom{2k-3}{k}}{4^k}, \tag{11}$$

which is defined for all $k \in \mathbb{Z}$ can be written as

$$t(k) = \begin{cases} 2s(k), k < 2, \\ s(k), \quad k \geq 2, \end{cases}$$

where

$$s(k) = (2 - k)\frac{(-1/2)_k}{4(1)_k}$$

is the proper sequence from Example 2.2. For $w \geq 1$, one should first split summation range in two

$$\sum_{k=0}^{w} t(k) = \frac{3}{4} + \sum_{k=2}^{w} s(k),$$

then the discrete Newton-Leibniz formula can be safely used to evaluate the sum on the right. However, applying directly (2) to (11) with (6) we obtain

$$\sum_{k=0}^{w} t(k) = (?) \quad u(w+1) - u(0) = \frac{(w+1)(w+2)\binom{2w-1}{w+1}}{2(w-1)4^w}. \tag{12}$$

If we assume that the value of $\binom{2k-3}{k}$ is 1 when $k = 0$ and -1 when $k = 1$ (that is natural from combinatorial point of view) then the expression on the right gives the true value of the sum only at $w = 0$.

2.3. *When the interval I contains no leading integer singularity of L*

Definition 2.3. For a linear difference operator (3) we call $M = \max(\{k \in \mathbb{Z}; a_1(k-1) = 0\} \cup \{-\infty\})$ the *maximal leading integer singularity of L*,

Proposition 2.1.[4] *Let $R(k)$ be a rational solution of (4). Then $R(k)$ has no poles larger than $M - 1$.*

Theorem 2.3.[4] *Let $L \in \mathcal{L}$, M be the maximal integer singularity of L, $l \geq M$, $I = \mathbb{Z}_{\geq l}$ and $t(k) \in V_I(L)$. Let Gosper's equation, corresponding to L, have a solution $R(k)$ in $K(k)$. Then $t(k) \in W_I(R(k), L)$.*

Example 2.4. For the sequence (11) we have $a_0(k) = -(2k-1)(k-1)$, $a_1(k) = 2(k+1)(k-2)$, $R(k) = 2k(k+1)/(k-2)$, and $u(k) = 2k(k+1)\binom{2k-3}{k}/((k-2)4^k)$. Thus $M = 3$, and the only pole of $R(k)$ is $k = 2$. As predicted by Theorem 2.3, the discrete Newton-Leibniz formula is valid when, e.g., $3 \le v \le w$.

3. The spaces $V_I(L)$ and $W_I(R(k), L)$

3.1. *The structure of $W_I(R(k), L)$*

Theorem 3.1.[5] *Let $L \in \mathcal{L}$ and Gosper's equation, corresponding to L, have a solution $R(k) \in K(k)$, $\mathrm{den}(R) = g(k)$. Then*

$$W_I(R(k), L) = g(k) \cdot V_I(\mathrm{pp}(L \circ g(k))),$$

where the operator $\mathrm{pp}(L \circ g(k))$ is computed by removing from $L \circ g$ the greatest common polynomial factor of its coefficients.

In addition, if $R = \frac{f(k)}{g(k)}$, $f(k) \perp g(k)$, then the space of the corresponding primitives of the elements of $W_I(R(k), L)$ can be described as $f(k) \cdot V_I(\mathrm{pp}(L \circ g(k)))$.

We will denote by $\bar{L}$ the operator $\mathrm{pp}(L \circ g(k))$.

Example 3.1. Consider again the operator $L = kE - (k+1)^2$ from Example 1.1 with $I = \mathbb{Z}$. We have $R = \frac{1}{k}$, and

$$L \circ k = kE \circ k - (k+1)^2 k = k(k+1)E - (k+1)^2 k = k(k+1)(E - k - 1),$$

$$\bar{L} = E - (k+1).$$

The space $W_I(R(k), \bar{L})$ is generated by $\bar{t}_2$, and, resp., the space $k \cdot W_I(R(k), \bar{L})$ is generated by $k\bar{t}_2$. In accordance with Theorem 3.1 the space $W_I(R(k), L)$ coincides with $k \cdot V_I(\bar{L})$.

It is possible to give examples showing that in some cases $\dim W_I(R(k), L) > 1$.

Example 3.2.
Let $L = 2(k^2 - 4)(k-9)E - (2k-3)(k-1)(k-8)$, $I = \mathbb{Z}$. Then Gosper's equation, corresponding to L, has the rational solution

$$R(k) = -\frac{2(k-3)(k+1)}{k-9}.$$

Here $g(k) = k - 9$ and $\bar{L} = 2(k^2 - 4)E - (2k-3)(k-1)$. Any sequence $\bar{t}$ which satisfies the equation $\bar{L}\bar{t} = 0$ has $\bar{t}(k) = 0$ for $k = 2$ or $k \le -2$. The values of $\bar{t}(1)$ and $\bar{t}(3)$ can be chosen arbitrarily, and all the other values are determined uniquely by the recurrence $2(k^2-4)\bar{t}(k+1) = (2k-3)(k-1)\bar{t}(k)$. Hence dim $V_I(\bar{L}) = 2$.

At the same time, dim $V_I(L) = 3$. Indeed, if $Lt = 0$, then $t(-2) = t(2) = t(9) = 0$. The value $t(k) = 0$ from $k = -2$ propagates to all $k \le -2$, but on each of the integer intervals $[-1, 0, 1]$, $[3, 4, 5, 6, 7, 8]$ and $[10, 11, \dots)$ we can choose one value arbitrarily, and the remaining values on that interval are then determined uniquely. A sequence $t(k) \in V_I(L)$ belongs to $W_I(R(k), L)$ iff $22t(10) - 13t(8) = 0$. So dim $W_I(R(k), L) = 2$.

3.2. When a rational solution of Gosper's equation is not unique

We give an example showing that if $L \in \mathcal{L}$ and Gosper's equation, corresponding to L, has different solutions $R_1(k), R_2(k) \in K(k)$, then it is possible that $W_I(R_1(k), L) \ne W_I(R_2(k), L)$. Moreover, these two spaces can have different dimensions.

Example 3.3. If $L = kE - (k+1)$, then Gosper's equation, corresponding to L, is

$$-(k+1)R(k+1) + kR(k) = -k,$$

and its general rational solution is

$$\frac{k-1}{2} + \frac{c}{k} = \frac{k^2 - k + 2c}{2k}.$$

Consider the solutions

$$R_1(k) = \frac{k-1}{2} \ (g_1(k) - 1), \quad \text{and} \quad R_2(k) - \frac{k^2 - k + 2}{2k} \ (g_2(k) - k).$$

We have $L \circ g_1(k) = L$, and $W_I(R_1(k), L) = V_I(L)$. This space has a basis that consists of two linearly independent sequences:

$$t_1(k) = \begin{cases} k, & \text{if } k \le 0, \\ 0, & \text{if } k > 0 \end{cases}$$

and

$$t_2(k) = \begin{cases} 0, & \text{if } k \le 0, \\ k, & \text{if } k > 0. \end{cases}$$

So this space contains, e.g., the sequence $t(k) = |k|$.

We have $L \circ g_2(k) = k(k+1)(E-1)$, therefore $W_I(R_2(k), L)$ is generated by the sequence $t(k) = k$.

If Gosper's equation, corresponding to $L \in \mathcal{L}$, has non-unique solution in $K(k)$, then the equation $Ly = 0$ has a non-zero solution in $K(k)$.

3.3. If Gosper's equation has a rational solution $R(k)$ then $W_I(R, L) \neq 0$

Theorem 3.2.[5] *Let* $L \in \mathcal{L}$ *and let Gosper's equation, corresponding to* L, *have a solution* $R(k) \in K(k)$. *Then* $W_I(R(k), L) \neq 0$ *(i.e.,* $\dim W_I(R(k), L) \geq 1$).

Example 3.4.

Let $L = (k+2)E - k$. The rational function $\frac{1}{k(k+1)}$ is a solution in $K(k)$ of the equation $Ly = 0$. Here $R(k) = -k - 1$, and $-1/k$ is a solution of the corresponding telescoping equation:

$$-\frac{1}{k+1} + \frac{1}{k} = \frac{1}{k(k+1)}.$$

The rational functions

$$\frac{1}{k(k+1)} \quad \text{and} \quad -\frac{1}{k}$$

have integer poles. Nevertheless, by Theorem 3.2 it has to be $W_I(R(k), L) \neq 0$ even when $I = \mathbb{Z}$. The space $W_I(R(k), L)$ is generated by the sequence

$$t(k) = \begin{cases} 1, & \text{if } k = -1, \\ -1, & \text{if } k = 0, \\ 0, & \text{otherwise,} \end{cases}$$

while the primitive of $t(k)$ is

$$(-k - 1)t(k) = \begin{cases} 1, & \text{if } k = 0, \\ 0, & \text{otherwise.} \end{cases}$$

If $I = \mathbb{Z}_{\geq 1}$, then $W_I(R(k), L)$ is generated by the sequence $t'(k) = \frac{1}{k(k+1)}$.

By Theorem 2.3, if M is the maximal integer singularity of L, $l \geq M$, $I = \mathbb{Z}_{\geq l}$, and Gosper's equation, corresponding to L, has a solution $R(k)$ in $K(k)$, then $V_I(L) = W_I(R(k), L)$. As a consequence, $\dim V_I(L) = \dim W_I(R(k), L) = 1$.

References

1. S. A. Abramov, On the summation of rational function, *USSR Comput. Math. Phys.* **11** (1971), 324–330. Transl. from *Zh. vychisl. mat. mat. fyz.* **11** (1971), 1071–1075.

2. S. A. Abramov, Rational solutions of linear difference and differential equations with polynomial coefficients, *USSR Comput. Math. Phys.* **29** (1989), 7–12. Transl. from *Zh. vychisl. mat. mat. fyz.* **29** (1989), 1611–1620.

3. S. A. Abramov, Rational solutions of linear difference and q-difference equations with polynomial coefficients, *Programming and Comput. Software* **21** (1995), 273–278. Transl. from *Programmirovanie* **21** (1995), 3–11.

4. S. A. Abramov and M. Petkovšek, Gosper's Algorithm, Accurate Summation, and the discrete Newton-Leibniz formula, ISSAC'05 (Annual International Symposium on Symbolic and Algebraic Computation). Beijing, China; 24–27 July 2005; *Proceedings ISSAC'05*, 5–12.

5. S. A. Abramov, On the summation of P-recursive sequences, ISSAC'06 (Annual International Symposium on Symbolic and Algebraic Computation). Genova, Italy; 9–12 July 2006; *Proceedings ISSAC'06*, 17–22.

6. R. W. Gosper, Jr., Decision procedure for indefinite hypergeometric summation, *Proc. Natl. Acad. Sci. USA* **75** (1978), 40–42.

FIVE APPLICATIONS OF WILF-ZEILBERGER THEORY TO ENUMERATION AND PROBABILITY *

MOA APAGODU

Department of Mathematics,
Virginia Commonwealth University,
Richmond, VA 23284-2014, USA.
E-mail: mapagodu at vcu dot edu.

DORON ZEILBERGER

Department of Mathematics,
Rutgers University (New Brunswick),
Hill Center-Busch Campus, 110 Frelinghuysen Rd.,
Piscataway, NJ 08854-8019, USA.
E-mail: zeilberg at math dot rutgers dot edu

*Dedicated to Sergei Abramov on becoming **FIVE**-dozen years old*

Explicit Formulas vs. Algorithms

In the old days, when one had to find some sequence, $a(n)$, there were two extremes. In the lucky case, one had an **explicit formula**. For example, the probability of tossing a fair coin $2n$ times and getting exactly n Heads, equals $(2n)!/(2^{2n}n!^2)$. Sometimes, cheatingly, one considered as 'explicit' expressions in terms of sums (or multisums) or integrals (or multi-integrals). The other extreme was to just have a *numerical* algorithm, that for each (numeric!) input n, found the output. In that case the algorithm was rated by its efficiency. Another compromise was an **asymptotic** formula, valid (approximately!) for large n.

But what's a formula?, it is a kind of algorithm. Of course, it is more than that, theoretically, but from a practical point of view it should be judged by the efficiency of the implied algorithm.

---

*Accompanied by Maple packages `AppsWZ` and `AppsWZmulti` downloadable from Zeilberger's website. Sample input and output can be viewed in:
`http://www.math.rutgers.edu/~zeilberg/mamarim/mamarimhtml/appswz.html`. The research of the second author was supported in part by the NSF.

The Holonomic Ansatz

Let's look at the explicit formulas that are called 'closed-form', or more precisely *hypergeometric* sequences. A sequence $a(n)$ is called hypergeometric if the ratio $a(n+1)/a(n)$ is a *rational* function of n, i.e. a quotient $P(n)/Q(n)$ where $P(n)$ and $Q(n)$ are polynomials. For example for the above-mentioned probability of getting exactly n Heads when tossing a fair coin $2n$ times, $p(n) := (2n)!/(2^{2n}n!^2)$, we have $p(n+1)/p(n) = (2n+1)/(2(n+1))$, or, by cross-multiplying

$$2(n+1)p(n+1) - (2n+1)p(n) = 0.$$

This is an example of a *first*-order **linear recurrence equation with polynomial coefficients**. Once you have the trivial value $p(0) = 1$ you can use it to compile a table of $p(n)$ for $n < N$, for any desired N in $O(N)$ operations.

The same is true for solutions of any linear recurrence equation with polynomial coefficients,

$$\sum_{i=0}^{L} a_i(n)p(n+i) = 0,$$

of order L. The only difference is that we need L *initial conditions*, $p(0), p(1), \ldots, p(L-1)$. We also assume that $a_L(n) = 0$ has no positive integer roots.

Such sequences were dubbed *P-recursive* by Richard Stanley in his seminal paper,[5] but we prefer the name *holonomic*. Zeilberger[8] famously showed that many sequences that arise in combinatorics, probability, and elsewhere are holonomic, and this was made into a full-fledged algorithmic theory by Wilf and Zeilberger.[6]

In our humble opinion, a holonomic representation of a sequence is to be considered *explicit*, since it is almost as good as a *closed-form* (i.e. hypergeometric). In the previous literature on WZ theory, there were few scattered examples of potential applications, but the focus was on the theory and the algorithms, not on specific applications.

Why this Paper?

The purpose of this paper is to fill this gap. We only list five such applications, but the reader can doubtless find many others. It is hoped that our implementation of these five applications will aid the reader to implement other ones that he or she might be interested in.

The Maple packages AppsWZ and AppsWZmulti

This article is accompanied by two Maple packages. `AppsWZ` that does applications of the single-variable case (see Ref. 1), and `AppsWZmulti` that does applications of the multi-variable case (see Refs. 2,6).

Asymptotics

Another nice feature of being a solution of a linear recurrence equation with polynomial coefficients is that using the Birkhoff-Trijinski method (see Ref. 7 for a lucid exposition), one can deduce the *asymptotics* to any desired order. This algorithm has been implemented by us in Maple, and is part of both packages.

First Application: Rolling a Die

If instead of tossing a coin n times, you roll a k-faced die, marked with positive or negative numbers, and you win the amount shown on the landed face (or lose, if it is a negative number). What is the probability that after n rolls, you break even? More generally, how likely are you to win exactly d dollars?

If the ith face ($i = 1 \ldots k$) shows the amount m_i, and lands with probability p_i, let the *probability generating function* of the die be defined by

$$P(x) := \sum_{i=1}^{k} p_i x^{m_i}.$$

It is *very* well known (and *very very* easy to see), that the probability, let's call it $a_d(n)$, of winding up with d dollars after n rolls is the constant term of

$$\frac{P(x)^n}{x^d}.$$

But that's exactly grist for the Almkvist-Zeilberger mill! In the package **AppsWZ** this is accompanied by the commands `RecProbVisit` and `RecProbVisitE`.

For example, let $a(n)$ be the probability of rolling a fair (standard) die $2n$ times and having the total score being *exactly* the expected value $7n$. Then $a(n)$ satisfies the following third-order linear recurrence equation with polynomial coefficients

$$-4(5+2n)(2n+3)(2n+1)(7n+19)(5n+11)(7n+20)(7n+13)(n+2)(n+1)a(n)$$

$$+4(7n+20)(5+2n)(2n+3)(n+2)$$

$$(25480n^5 + 223496n^4 + 755066n^3 + 1223233n^2 + 946889n + 279936)a(n+1) -$$

$$(5n+6)(5+2n)(6+7n)(499359n^6 + 6777015n^5 + 38079431n^4$$

$$+113390385n^3 + 188723986n^2 + 166469280n + 60800544)a(n+2) +$$

$$30(5n+14)(5n+13)(5n+12)(7n+12)$$

$$\cdot(5n+11)(5n+6)(7n+13)(6+7n)(n+3)a(n+3) = 0,$$

and the Birkhoff-Trijinksi method implies that the asymptotics is:

$$(.197833497170804)n^{-1/2}(1 - (111/1400)/n - (12037/5488000)/n^2$$

$$+(1367631/1097600000)/n^3 + \ldots).$$

Readers can produce their own output for the scenario of their choice.

Second Application: How many ways to have r people chip in to pay a bill of n cents

In George Pólya's classic 'picture writing' paper,[4] he considers the problem of figuring out how many ways can *one* person pay a bill of n cents using any number of coins. If the denominations are $\{d_1, \ldots, d_k\}$ ($\{1, 5, 10, 25, 50, 100\}$ in the US), then the required number is the coefficient of x^n in the *generating function*

$$\prod_{i=1}^{k} \frac{1}{1 - x^{d_i}}.$$

Calling this number $a(n)$, this entails, trivially, a *linear recurrence equation with constant coefficients of order* $\sum_{i=1}^{k} d_i$. But by allowing **polynomial** coefficients, one can get, thanks to Almkvist-Zeilberger, a recurrence of order $\leq lcm(d_1, \ldots, d_k)$.

More generally, of $a_r(n)$ is the number of ways of breaking n cents with (up to) r people participating (or equivalently, one person with r pockets in his or her pants or dress), the generating function is

$$\prod_{i=1}^{k} \frac{1}{(1 - x^{d_i})^r},$$

and applying Almkvist-Zeilberger to

$$\frac{1}{x^{n+1}} \prod_{i=1}^{k} \frac{1}{(1 - x^{d_i})^r},$$

produces a linear recurrence, still with polynomial coefficients (but now these are polynomials in both n and r) of the above order. On the other hand if you stick to *constant* coefficients then the order would be $r(\sum_{i=1}^k d_i)$, and would only make sense for (small!!) *numeric r*, while within the *holonomic ansatz*, one can have *symbolic r* without any increase in the order. The procedure that takes care of this problem in the Maple package AppsWZ is TeamEffortMoneyChanging. For example, entering:

TeamEffortMoneyChanging({1,5,10,25},n,N,r);

would tell you that if $a_r(n)$ is the number of ways of breaking n cents with (up to) r people chipping-in, using only pennies, nickels, dimes, and quarters, and denoting by N the shift operator in n: $(Nx(n) := x(n+1))$, it turns out that $a_r(n)$ is annihilated by the following 30th-order linear recurrence operator with polynomial coefficients:

$$(-41r - n) + (-r)N + (-r)N^2 + (-r)N^3 + (-r)N^4 +$$

$$(-47r - n - 5)N^5 + (-2r)N^6 + (-2r)N^7 + (-2r)N^8 + (-2r)N^9 + (-22r)N^{10}$$

$$+(-2r)N^{11} + (-2r)N^{12} + (-2r)N^{13} + (-2r)N^{14}$$

$$+(-22r)N^{15} + (-2r)N^{16} + (-2r)N^{17} + (-2r)N^{18} + (-2r)N^{19} + (-22r)N^{20} +$$

$$(-2r)N^{21} + (-2r)N^{22} + (-2r)N^{23} + (-2r)N^{24} + (-6r + n + 25)N^{25}$$

$$+(-r)N^{26} + (-r)N^{27} + (-r)N^{28} + (-r)N^{29} + (n + 30)N^{30}.$$

Third Application: Hidden Markov Models

In contemporary bioinformatics (see for example the fascinating and lucid expository article by Lior Pachter and Bernd Sturmfels in the wonderful book that they edited(see Ref. 3), one has to *estimate* probabilities from output. The usual approach is *maximum likelihood*, that entails solving *huge* systems of polynomial equations that are handled via Buchberger's Gröbner Basis Algorithm. But, following Laplace, we can also use a *Bayesian* approach and get quotients of integrals. These integrals can be handled via the multi-Almkvist-Zeilberger (see Ref. 2) algorithm. For any *specific* instance one may want to use *numerical integration*. If one wants to compile a table anticipating all possible outputs (of course up to a certain length), then the recurrences supplied by Almkvist-Zeilberger are much more efficient.

Let's illustrate it with the simple case of (only) two *dice*, that enables us to stay within the single-variable Almkvist-Zeilberger (see Ref. 1).

In the proverbial *dishonest casino*, suppose that there are two kinds of (identically looking) dice, with the same possible outcomes (faces), let's call them $\{1, \ldots, m\}$, each with a *known* probability distribution $(p_1, \ldots, p_m)$ and $(q_1, \ldots, q_m)$. What is unknown is the probability, x, of using the first die (and hence $1 - x$ for the other die). Judging from the sequence of outcomes (or rather by the relative frequencies of the landed faces), one has to *estimate* x. If the (hidden) probability of using the first die was x, (whatever it is), then the probability of the output distribution being $a_1, \ldots, a_m$ is $(a_1 + \cdots + a_m)!/(a_1! \cdots a_m!)$ times

$$L(x) := \prod_{i=1}^{m} (xp_i + (1 - x)q_i)^{a_i}.$$

The *maximum-likelihood* estimate is to *maximize* $L(x)$ by solving $L'(x) = 0$ (with more dice one gets partial derivatives and systems of equations in several variables). But, following the more democratic approach of Laplace, that considers *all scenarios* (and that famously tells you that if so far you succeeded m times and failed n times, then your estimated probability of success in your next try is not $m/(m+n)$ but rather $(m+1)/(m+n+2)$), we would have not the root of $L'(x) = 0$ but rather

$$\frac{\int_0^1 x \prod_{i=1}^{m} (xp_i + (1 - x)q_i)^{a_i} \, dx}{\int_0^1 \prod_{i=1}^{m} (xp_i + (1 - x)q_i)^{a_i} \, dx}.$$

Calling the top quantity $T(a_1, \ldots, a_m)$ and the bottom quantity $B(a_1, \ldots, a_m)$, we see that Almkvist-Zeilberger can manufacture *pure* linear recurrences (with polynomial coefficients) in *each* of the discrete variables $a_1, \ldots, a_m$ (in Nature $m = 4$, or $m = 20$). The multi-dice analog of this (which would employ the multi-variable Almkvist-Zellberger, done in Ref. 2) may be of interest to bioinformaticians. So far we have only implemented the interface in the two-dice case.

The procedure that takes care of this problem in `AppsWZ` is `ExpProbFirstDie(Lp1,Lp2,a0)`, where `Lp1`, and `LP2` are the probability distributions and `a0` is the list of respective outcome. The novelty here is that both top and bottom integrals are not computed directly but via the recurrences outputted by Almkvist-Zeilberger. With the *option remember* it should be much more efficient if one wants to pre-compute a table of estimated probability for each `a0` of size less than some (lagre) pre-assigned value.

For example, if there are two coins, one fair and one loaded with probability of a Head being $1/3$, and the outcome was 10 and 10 (which ML would say

that the fair coin was used all the time),
`evalf(ExpProbFirstDie([1/2,1/2],[1/3,2/3],[10,10])):`
would estimate that the fair coin was used only %58.76 of the time.

Fourth Application: Lattice Paths Counting

We all know that the number of ways of walking from the origin $(0,0)$ to the point (m,n) in the square lattice, with unit northbound and unit eastbound steps $\{(0,1),(1,0)\}$ is

$$F(m,n) = \frac{(m+n)!}{m!n!}.$$

This immediately implies (and is equivalent to) the fact that

$$\frac{F(m+1,n)}{F(m,n)} = \frac{m+n+1}{m+1}, \quad \frac{F(m,n+1)}{F(m,n)} = \frac{m+n+1}{n+1}.$$

and cross-multiplying yields

$$(m+1)F(m+1,n) - (m+n+1)F(m,n) = 0,$$

$$(n+1)F(m,n+1) - (m+n+1)F(m,n) = 0.$$

In other words the discrete function $F(m,n)$ satisfies *pure* linear recurrences equations with *polynomial* coefficients, that happen, in this simple case, to be *first-order*. Recall that a recurrence is *pure* if only one of the variables changes at a time. For example, $F(m,n)$ trivially satisfies the "mixed" (*partial*) recurrence $F(m,n) = F(m-1,n) + F(m,n-1)$.

An amazing consequence of Wilf-Zeilberger theory[6] is that this is still true for an *arbitrary* set of (positive) steps, and in *arbitrary* dimension. Of course the pure recurrences are no longer (usually) first-order, but as above, this is a minor computational disadvantage.

Indeed, if we are walking in the d-dimensional (hyper)cubic lattice, starting at the origin, and with a set of steps S (all with non-negative coordinates, excluding the step $\mathbf{0}$ [staying in place]), the generating function is trivially seen to be

$$\sum_m F(m_1,\ldots,m_d){x_1}^{m_1}\cdots {x_d}^{m_d}$$

$$= \frac{1}{\left(1 - \sum_{(s_1,\ldots,s_d)\in S} {x_1}^{s_1}\cdots {x_d}^{s_d}\right)}.$$

So it follows that our discrete function of interest, $F(m_1, \ldots, m_d)$ equals the formal residue of

$$\frac{x_1^{-m_1-1} \cdots x_d^{-m_d-1}}{\left(1 - \sum_{(s_1,\ldots,s_d) \in S} x_1^{s_1} \cdots x_d^{s_d}\right)}.$$

If you are only interested in getting to points on the diagonal, then $f(n) := F(n, n, \ldots, n)$ is given by the formal residue of

$$\frac{x_1^{-n-1} \cdots x_d^{-n-1}}{\left(1 - \sum_{(s_1,\ldots,s_d) \in S} x_1^{s_1} \cdots x_d^{s_d}\right)},$$

and once again thanks to Wilf-Zeilberger theory, satisfies a linear recurrence equation with polynomial coefficients.

This is implemented in the second Maple package accompanying this paper, `AppsWZmulti` that is "powered" by the Maple package `MultiAlmkvistZeilberger` that accompanied.[2] The relevant procedures (in `AppsWZmulti`) are `LatticePaths` and `LatticePathsDiagonal` and for the verbose versions `LatticePathsStory` and `LatticePathsDiagonalStory`.

For example, if you type `LatticePaths({[0,1],[1,0],[1,1]},m,M);` you would get that the following two operators annihilate $F(m_1, m_2)$, the number of ways of going from $(0,0)$ to (m_1, m_2) using the steps $\{(0,1),(1,0),(1,1)\}$ (where M_1, M_2 are the shift operators in the m_1, m_2 variables respectively)

$$[-(m_1+1)/(2+m_1) - ((1+2m_2)/(2+m_1))M_1 + M_1^2,$$

$$-(m_2+1)/(2+m_2) - ((1+2m_1)/(2+m_2))M_2 + M_2^2],$$

which in everyday parlance means that $F(m_1, m_2)$ satisfies

$$(m_1+2)F(m_1+2, m_2) - (2m_2+1)F(m_1+1, m_2) - (m_1+1)F(m_1, m_2) = 0,$$

$$(m_2+2)F(m_1, m_2+2) - (2m_1+1)F(m_1, m_2+1) - (m_2+1)F(m_1, m_2) = 0.$$

If you type `LatticePathsDiagonal({[0,1],[1,0],[1,1]},n,N);` you would get that $f(n)$, the number of ways of getting from $(0,0)$ to (n,n) using the same set of steps is:

$$(n+2)f(n+2) - 3(2n+3)f(n+1) + (n+1)f(n) = 0,$$

subject to the initial conditions $f(0) = 1, f(1) = 3$. Thanks to Birkhoff-Trijinski, its asymptotics is

$$C(3 + 2^{3/2})^n n^{-1/2} (1 + ((3/32)2^{1/2} - 1/4)/n + (113/1024 - (9/128)2^{1/2})/n^2$$

$$+((1545/32768)2^{1/2} - 245/4096)/n^3),$$

for some constant C.

A more interesting example are the (old-time) **basketball numbers**, which is the number of ways a basketball game that ended with the score $n : n$ can proceed. Recall that in the old days (before 1961), an atom of basketball-scoring could be only of one or two points. Equivalently, this number is the number of ways of walking, in the square lattice, from $(0,0)$ to (n, n) using the atomic steps $\{(1,0), (2,0), (0,1), (0,2)\}$. Entering this into LatticePathsDiagonalStory yields that, calling this number $F(n)$, it satisfies the third-order linear recurrence:

$$(16/5)(2n + 3)(11n + 26)(1 + n)/((n + 3)(2 + n)(11n + 15))F(n)$$

$$-(4/5)(121n^3 + 649n^2 + 1135n + 646)/((n + 3)(2 + n)(11n + 15))F(1 + n)$$

$$-(2/5)(176n^2 + 680n + 605)/((11n + 15)(n + 3))F(2 + n) + F(n + 3) = 0,$$

subject to the initial conditions:

$$F(0) = 1, F(1) = 2, F(2) = 14.$$

For the record, the first few terms are:
[1, 2, 14, 84, 556, 3736, 25612, 177688, 1244398, 8777612, 62271384, 443847648, 3175924636, 22799963576, 164142004184, 1184574592592, 8567000931404, 62073936511496, 450518481039956, 3274628801768744, 23833760489660324].

The asymptotics is:

$$(.37305616)(4 + 2(3^{1/2}))^n n^{-1/2}(1 + (67/1452)3^{1/2} - (119/484))/n$$

$$+((6253/117128) - (7163/234256)3^{1/2})/n^2$$

$$+(-(32645/15460896)3^{1/2} + (129625/10307264))/n^3),$$

or in floating-point:

$$(.37305616)(7.464101616)^n n^{-1/2} \cdot$$

$$(1. - .1659453140/n + .42398086 \cdot 10^{-3}/n^2 + .8918933381 \cdot 10^{-2}/n^3).$$

Fifth Application: Random Walk in Higher Dimensions

This is the multivariate analog of the First Application. The relevant procedures in AppsWZmulti are RandomWalkRecurrence and

RandomWalkRecurrenceE. By now readers should be able to generate their own examples. A few sample input and output files are given in the webpage of this article.

References

1. G. Almkvist and D. Zeilberger, *The method of differentiating under the integral sign*, J. Symbolic Computation **10**, 571-591 (1990).
2. M. Apagodu and D. Zeilberger *Multi-Variable Zeilberger and Almkvist-Zeilberger Algorithms and the Sharpening of Wilf-Zeilberger Theory* , Adv. Appl. Math. **37** (2006), 139-152.
3. L. Pachter and B. Sturmfels *Statistics*, in: *"Algebraic Statistics for Computational Biology"* [L. Pachter and B. Sturmfels, eds.], Cambridge University Press, 2005, 3-33.
4. G. Pólya *On picture writing*, Amer. Math. Monthly **63** (1956), 689-697.
5. R. Stanley *Differentiably finite power series*, Europ. J. Combinatorics **1** (1980), 175-188.
6. H.S. Wilf and D. Zeilberger *An algorithmic proof theory for hypergeometric (ordinary and "q") multisum/integral identities*, Invent. Math. **108**(1992), 575-633 .
7. J. Wimp and D. Zeilberger *Resurrecting the asymptotics of linear recurrences*, J. Math. Anal. Appl. **111**(1985), 162-177 .
8. D. Zeilberger *A Holonomic systems approach to special functions identities*, J. of Computational and Applied Math. **32** (1990), 321-368 .

FACTORING SYSTEMS OF LINEAR FUNCTIONAL
EQUATIONS USING EIGENRINGS *

MOULAY A. BARKATOU

*Institut XLIM UMR 6172 CNRS, Université de Limoges,
123, Avenue Albert Thomas, 87060 Limoges Cedex, France
* E-mail: Moulay.Barkatou@unilim.fr*

We present a simple method for computing factorizations for a large class of matrix equations (matrix pseudo-linear equations) that includes, in particular, systems of linear differential, difference and $q-$difference equations. The approach is based on the characterization of the properties (reducibility, decomposability, etc.) of matrix pseudo-linear equations in terms of their eigenrings.

Keywords: pseudo-linear equations, differential equations, difference equations, $q-$difference equations, reducibility, decomposability, eigenrings, computer algebra.

1. Introduction and notations

Throughout this paper, we let K be a commutative field of characteristic zero, ϕ an automorphism of K, δ a $\phi-$derivation, that is a map $\delta : K \to K$ satisfying

$$\delta(a + b) = \delta a + \delta b \text{ and } \delta(ab) = \delta a\, \phi b + a\, \delta b \text{ for all } a, b \in K.$$

Note that when $\phi = 1_K$ the identity map of K, δ is simply a derivation; otherwise, one can easily show that, δ has to be of the form $\gamma(1_K - \phi)$ for some $\gamma \in K$.
By C we denote the *field of constants* of (K, ϕ, δ)

$$C = \{c \in K \mid \phi c = c \text{ and } \delta c = 0\}.$$

If M is a matrix (vector) in K, we define ϕM (*resp.* δM) to be the matrix (vector) obtained by applying ϕ (*resp.* δ) to all components of M. We note

---

*This paper is a slightly expanded version of an unpublished paper[7] that goes back to 2001. The results have been extended to the case of positive characteristic in Refs. 10 and 4.

that the operations on matrices (vectors) commute with ϕ. If M, N are matrices in K then

$$\delta(M + N) = \delta M + \delta N, \quad \delta(MN) = \delta M \phi N + M \delta N.$$

By a matrix *pseudo-linear* equation over (K, ϕ, δ) we mean an equation of the form

$$\delta y = -A\phi y \tag{1}$$

where A is an $n \times n$ matrix in K and where y is an unknown $n-$dimensional column vector. A matrix pseudo-linear equation (1) will be denoted by $[A]$. We note the following well-known examples of matrix pseudo-linear equations

(a) Differential equation: $K = \mathbb{C}(x)$ or $\mathbb{C}((x))$, $\phi = 1_K$ and $\delta = \frac{d}{dx}$.
(b) Difference equation: $K = \mathbb{C}(x)$ or $\mathbb{C}((x^{-1}))$, ϕ the $\mathbb{C}-$automorphism defined by $\phi x = x - 1$ and $\delta = 1 - \phi$.
(c) $q-$Difference equation: $K = \mathbb{C}(x)$ or $\mathbb{C}((x^{-1}))$, ϕ the $\mathbb{C}-$automorphism defined by $\phi x = qx$, $q \in \mathbb{C}^*$ and $\delta = 1 - \phi$.

Consider a matrix equation (1) and let $P \in \mathrm{GL}(n, K)$. The substitution $y = Pz$ leads to the matrix equation

$$\delta z = -(P^{-1}A\phi P + P^{-1}\delta P)\phi z.$$

We say that two $n \times n$ matrices A, B in K are *equivalent* (notation: $A \sim B$) if there exists a matrix $P \in \mathrm{GL}(n, K)$ such that

$$B = P^{-1}A\phi P + P^{-1}\delta P.$$

We say that the pseudo-linear equations $[A]$ and $[B]$ are equivalent if their matrices are equivalent.

A matrix pseudo-linear equation $[A]$ is said to be *reducible*, if A is equivalent to a block–triangular matrix

$$B = \begin{pmatrix} B_{1,1} & 0 \\ B_{2,1} & B_{2,2} \end{pmatrix} \tag{2}$$

where $B_{i,i}$, $i = 1, 2$, is a square matrix of size $n_i < n$. It is called *decomposable* if A is equivalent to a matrix of the form (2) with $B_{2,1} = 0$. It is called *irreducible (indecomposable)* if it is not reducible (decomposable). It is said to be *completely reducible*, if A is equivalent to a block–diagonal matrix

$$B = B_{1,1} \oplus \ldots \oplus B_{s,s} \stackrel{def}{=} \begin{pmatrix} B_{1,1} & & 0 \\ & \ddots & \\ 0 & & B_{s,s} \end{pmatrix} \tag{3}$$

where each equation $[B_{i,i}]$, $1 \leq i \leq s$, is irreducible. In particular, an equation which is irreducible is completely reducible (take $s = 1$). Let us note that an equation which is completely reducible and reducible is necessarily decomposable.

The *eigenring* of the matrix pseudo-linear equation $[A]$ is defined as the set of $n \times n$ matrices T in K satisfying

$$\delta T = TA - A\phi(T).$$

We shall denote by $\mathcal{E}(A)$ the eigenring of $[A]$. It is a $\mathcal{C}$−algebra which has finite dimension, as vector space over $\mathcal{C}$ (cf. Sec. 2). Algorithms for computing eigenrings of a differential (difference) systems over $\mathbb{C}(x)$ have been presented in Refs. 8 and 6.

Our interest in eigenrings arises from the fact we shall establish later that with their aid we can *factor* matrix pseudo-linear equations. By *to factor* a pseudo-linear equation, we mean to compute an equivalent equation having a block-triangular or block-diagonal form or to show that such an equation does not exist.

The problem of testing reducibility of general matrix pseudo-linear equation seems to be difficult and will not be discussed here. The reader who is interested in this problem is referred to Refs. 12,15,16 for a discussion of this problem in the differential case. When one considers restricted classes of matrix pseudo-linear equations, the problem of factorization becomes more tractable. In this paper we shall consider the class of matrix pseudo-linear equations the eigenring of which is not a division ring. Our main results are exposed in Section 3. We shall show that equations in this class are reducible and give an elementary method to factor them. This class contains, in particular, matrix pseudo-linear equations that are decomposable. We shall see that a matrix pseudo-linear equation $[A]$ is decomposable if, and only if, its eigenring contains a matrix T which has at least two distinct eigenvalues (see Theorem 3.2). Furthermore, the equation can be factored by finding a matrix $P \in GL(n, K)$ such that $P^{-1}TP$ have a block-diagonal form. This yields a simple algorithm that, given A and $\mathcal{E}(A)$, determines if the equation $[A]$ is decomposable and, if the case arises, computes an equivalent equation $[B]$ of the form (3) where the $[B_{i,i}]$ are indecomposable. In particular, this algorithm allows to factor equations that are completely reducible and reducible.

The idea of using eigenrings in problems of factorization has been introduced by Singer.[16] It has been used by M. van Höij to factor scalar differential equations.[17] We have also already apply this idea to compute factorizations of systems of linear differential (difference) equations.[6,8] However, what is new is that, in one hand, the results in the present paper are more complete than those in Refs. 8 and 6 and improve them (for example Theorem 3.2 was not in Refs. 6,8); on the other hand they are presented in an unifying setting that permits us to treat simultaneously a large class of matrix (or scalar) linear functional equations.

The rest of the paper is organized as follows. In Section 2 we list various facts about matrix pseudo-linear equations that will be used in the others sections. In Section 4 we generalize some of the results of Section 3. With two matrix pseudo-linear equations $[A]$, $[B]$ of dimension m, n we associate the set $\mathcal{H}(A, B)$ of $n \times m$ matrices T in K satisfying the equation

$$TA - B\phi(T) = \delta T.$$

It is a $\mathcal{C}$−vector space of finite dimension. When $A = B$, one has $\mathcal{H}(A, B) = \mathcal{E}(A)$. We show how in a number of cases the equations $[A]$ and $[B]$ can be factored using $\mathcal{H}(A, B)$.

For the sake of completeness we have added an appendix on the relationship between matrix pseudo-linear equations and modules on the Ore polynomial ring $K[X; \phi, \delta]$. The language of modules throws a different light on matrix pseudo-linear equations. However, the results we shall list in the appendix will not be used in the first part of the paper. Using modules language the proofs of some of our results though elementary could be further simplified. However, since we are interested in algorithmic aspects, we prefer to state and prove our results in matrix terms.

2. Preliminaries

In this section we list a few facts about matrix pseudo linear equations that will be used in later sections.

Consider a matrix pseudo-linear equation $[A]$ over (K, ϕ, δ).
The *solution space over K* of $[A]$ is defined as the set

$$Sol_K(A) = \{v \in K^n | \delta v + A\phi v = 0\}.$$

We will show that it is a finite dimensional vector space over the field of constants $\mathcal{C}$. For this we need first the following

Lemma 2.1. *If the vectors* $v_1, \ldots, v_m \in Sol_K(A)$ *are linearly dependent over K then they are linearly dependent over $\mathcal{C}$.*

Proof. We procced by induction on m. The case $m = 1$ is trivial. Let $m > 1$ and let $v_1, \ldots, v_m \in Sol_K(A)$ be linearly dependent over K. We may suppose that any proper subset of $\{v_1, \ldots, v_m\}$ is linearly independent over K. Then there is a unique relation

$$v_1 = \sum_{i=2}^{m} a_i v_i$$

with all a_i in K. Now

$$0 = \delta v_1 + A\phi v_1 = \sum_{i=2}^{m} \phi a_i (\delta v_i + A\phi v_i) + \sum_{i=2}^{m} \delta a_i \, v_i = \sum_{i=2}^{m} \delta a_i \, v_i.$$

Thus all $\delta a_i = 0$. Now if $\phi = 1_K$ we are done, otherwise $\delta = \gamma(1_K - \phi)$ for some $\gamma \in K$ and hence $\phi a_i = a_i$ for all i. Thus all a_i belong to $\mathcal{C}$. $\square$

Proposition 2.1. $Sol_K(A)$ *is a vector space over the field of constants* $\mathcal{C}$ *of dimension* $\leq n$.

Proof. It is clear that $Sol_K(A)$ is a vector space over $\mathcal{C}$. Now any $n+1$ vectors in $Sol_K(A)$ are linearly dependent over K. According to the above lemma, they are also linearly dependent over $\mathcal{C}$. Hence the dimension of $Sol_K(A)$ over $\mathcal{C}$ is $\leq n$. $\square$

Let us note that algorithms for determining the space of solutions of a matrix differential (difference) equation over $\mathcal{C}(x)$ have been developed in Refs. 1,3,5,6,18. The case of matrix (q-)difference equations has been discussed in Ref. 2. The case of matrix differential equations over $K = \mathcal{C}((x))$ has been considered in Ref. 9. Some of these algorithms have been implemented in Maple.

The eigenring of A is the set $\mathcal{E}(A)$ of matrices T in K that are solutions of the equation

$$\delta T = TA - A\phi(T).$$

This equation can be viewed as a system of n^2 first-order pseudo-linear equations over (K, ϕ, δ). Thus, $\mathcal{E}(A)$ is a $\mathcal{C}-$vector space of finite dimension $\leq n^2$. On the other hand, a direct calculation shows that the product of two elements of $\mathcal{E}(A)$ is also an element of $\mathcal{E}(A)$. Finally, the identity matrix I_n belongs to $\mathcal{E}(A)$. Thus, $\mathcal{E}(A)$ is an algebra over $\mathcal{C}$ and as a $\mathcal{C}-$vector space it has dimension $\leq n^2$. As a consequence, we have the following

Proposition 2.2. *Any element* T *of* $\mathcal{E}(A)$ *has a minimal polynomial with coefficients in the field of constants* $\mathcal{C}$. *In particular, when* $\mathcal{C}$ *is algebraically closed, each matrix* $T \in \mathcal{E}(A)$ *has all its eigenvalues in* $\mathcal{C}$.

We conclude this section by the

Proposition 2.3. *If two equations* $[A]$ *and* $[B]$ *are equivalent, their eigen-rings* $\mathcal{E}(A)$ *and* $\mathcal{E}(B)$ *are isomorphic as* C $-$ *algebras. In particular, one has* $dim_C\mathcal{E}(A) = dim_C\mathcal{E}(B)$

Proof. Suppose that $A \sim B$ then there exists a matrix $P \in \mathrm{GL}(n, K)$ such that $B = P^{-1}A\phi P + P^{-1}\delta P$. A direct calculation shows that the map $T \longrightarrow P^{-1}TP$ defines an isomorphism from $\mathcal{E}(A)$ into $\mathcal{E}(B)$. $\square$

3. Eigenrings and reduction of pseudo-linear equations

We shall give some sufficient conditions for reducibility, decomposability which are easy to check. The proofs of the following propositions are elementary and give effective methods to reduce (decompose) a given reducible (decomposable) matrix pseudo-linear equation.

Throughout this section we let $[A]$ be a given matrix pseudo-linear equation and we suppose that its eigenring $\mathcal{E}(A)$ is known.

Theorem 3.1. *If* $\mathcal{E}(A)$ *is not a division ring then* $[A]$ *is reducible and the reduction can be carried out by a matrix* $P \in GL(n, K)$ *that can be computed explicitly.*

Proof. We have to prove that if $\mathcal{E}(A)$ is not a division ring, one can construct a transformation $P \in GL(n, K)$ such that

$$P^{-1}A\phi P + P^{-1}\delta P = \begin{pmatrix} B_{1,1} & 0 \\ B_{2,1} & B_{2,2} \end{pmatrix}$$

where $B_{1,1}$ and $B_{2,2}$ are square matrices of order $< n$.

Suppose that $\mathcal{E}(A)$ is not a division ring. Then one can choose an element $T \in \mathcal{E}(A) \setminus \{0\}$ which is not invertible. Let r denote the rank of T, then $0 < r < n$. Performing Gaussian elimination on the columns of T one can compute $P \in GL(n, K)$ such that the matrix defined by $S = P^{-1}TP$ be of the form

$$S = \begin{pmatrix} S_{1,1} & 0 \\ S_{2,1} & 0 \end{pmatrix},$$

where $S_{1,1}$ is an $r \times r$ matrix in K, $S_{2,1}$ an $(n - r) \times r$ matrix in K and where $\begin{pmatrix} S_{1,1} \\ S_{2,1} \end{pmatrix}$ has rank r. Let $B = P^{-1}(A\phi P + \delta P)$ then $S \in \mathcal{E}(B)$, that is S satisfies the equation $SB - B\phi S = \delta S$. We decompose B in the same form as S

$$B = \begin{pmatrix} B_{1,1} & B_{1,2} \\ B_{2,1} & B_{2,2} \end{pmatrix}.$$

Then the relation $SB - B\phi S = \delta S$ implies

$$\begin{pmatrix} S_{1,1} \\ S_{2,1} \end{pmatrix} B_{1,2} = 0.$$

Since $\begin{pmatrix} S_{1,1} \\ S_{2,1} \end{pmatrix}$ is of full rank, then $B_{1,2} = 0$. $\qquad\square$

The proof of the following corollary is immediate.

Corollary 3.1. *Given a matrix pseudo-linear equation $[A]$ one can construct an equivalent matrix equation $[B]$ having a block-triangular form*

$$\begin{pmatrix} B_{1,1} & 0 & & 0 \\ B_{2,1} & B_{2,2} & & \\ \vdots & & \ddots & 0 \\ B_{k,1} & \cdots & & B_{k,k} \end{pmatrix}$$

where k is the maximal possible, i.e. for each $1 \le i \le k$, the eigenring of the $n_i \times n_i$ matrix $B_{i,i}$ is a division ring.

Let us note that the condition "$\mathcal{E}(A)$ is not a division ring" implies $\dim_{\mathcal{C}} \mathcal{E}(A) > 1$. Indeed, if $T \in \mathcal{E}(A) \setminus \{0\}$ is not invertible, then the family $\{I, T\}$ is linearly independent (over $\mathcal{C}$) and hence $\dim_{\mathcal{C}} \mathcal{E}(A) > 1$. The converse is true when the field of constants $\mathcal{C}$ is algebraically closed. Indeed, suppose that $\dim_{\mathcal{C}} \mathcal{E}(A) > 1$ then there exists $T \in \mathcal{E}(A)$ such that the family $\{I, T\}$ be linearly independent. Since $\mathcal{C}$ is algebraically closed, there exists $\lambda \in \mathcal{C}$ such that $\det(T - \lambda I) = 0$. Hence $\mathcal{E}(A)$ contains an element, namely $T - \lambda I$, which is non-zero and non invertible.

To check the condition "$\mathcal{E}(A)$ is not a division ring", in the case where the field of constants $\mathcal{C}$ is not algebraically closed, one can proceed as follows. Find a basis $\{T_1, \ldots, T_d\}$ of $\mathcal{E}(A)$, and decide if there are constants $c_i \in \mathcal{C}$ such that $(c_1, \ldots, c_d) \ne (0, \ldots, 0)$ and $\det\left(\sum_{i=1}^{d} c_i T_i\right) = 0$.

Example 3.1. Let $K = \mathbb{Q}(x)$, ϕ the automorphism of K defined by $\phi x = x - 1$ and $\delta = 1 - \phi$. Consider the difference system $[A]$ where A is given by

$$
A = \begin{bmatrix}
-(x-1)^{-1} - x + 2 & -2 & \frac{-3x-2+x^2}{x-1} \\[2mm]
(x-1)^{-1} & 0 & 1 & 3(x-1)^{-1} \\[2mm]
-\frac{1+x^2-x}{x(x-1)} & -\frac{x-2}{x} & -\frac{x^2-3x+2}{x} & -\frac{3+x^2-x}{x(x-1)} \\[2mm]
(x-1)^{-1} & x-2 & 1 & -\frac{-1+x^2-3x}{x-1}
\end{bmatrix}
$$

The eigenring of $[A]$ is the set of matrices T of the form

$$
T = \begin{bmatrix}
c_1 - 4c_2 & 0 & 0 & -4c_2 \\[2mm]
2c_2 & c_1 & 0 & 2c_2 \\[2mm]
-2\frac{c_2}{x} & 0 & c_1 - 2c_2 & -4\frac{c_2}{x} \\[2mm]
2c_2 & 0 & 0 & 2c_2 + c_1
\end{bmatrix}
$$

where c_1 and c_2 are arbitrary constants.
Here, $\det T = c_1^2(c_1 - 2c_2)^2$. Choose $c_1 = 0$ and $c_2 = 1$, then

$$
T = \begin{bmatrix}
-4 & 0 & 0 & -4 \\[2mm]
2 & 0 & 0 & 2 \\[2mm]
-2x^{-1} & 0 & -2 & -4x^{-1} \\[2mm]
2 & 0 & 0 & 2
\end{bmatrix}.
$$

One has $\operatorname{rank} T = 2$. So, performing Gaussian elimination on the columns of T, one finds

$$
P = \begin{bmatrix}
1 & 0 & -1 & 0 \\[2mm]
0 & 0 & 0 & 1 \\[2mm]
0 & 1 & -\frac{1}{x} & 0 \\[2mm]
0 & 0 & 1 & 0
\end{bmatrix}
$$

such that the matrix

$$
TP = \begin{bmatrix}
-4 & 0 & 0 & 0 \\[2mm]
2 & 0 & 0 & 0 \\[2mm]
-2\frac{1}{x} & -2 & 0 & 0 \\[2mm]
2 & 0 & 0 & 0
\end{bmatrix}
$$

have its two first columns linearly independent. Now if we compute the matrix $B = P^{-1}(A\phi P + \delta P)$ we get

$$B = \begin{bmatrix} 0 & -1 & 0 & 0 \\ -1 & -\frac{1+x^2-3x}{x} & 0 & 0 \\ (x-1)^{-1} & 1 & -\frac{1+x^2-3x}{x-1} & x-2 \\ (x-1)^{-1} & 1 & (x-1)^{-1} & 0 \end{bmatrix}.$$

Theorem 3.1 and its corollary give a method which allows to reduce a matrix equation $[A]$ whose eigenring $\mathcal{E}(A)$ is not a division ring. The following proposition gives a sufficient condition for that $\mathcal{E}(A)$ not be a division ring.

Proposition 3.1. *If $[A]$ is decomposable then $\mathcal{E}(A)$ is not a division ring.*

Proof. If $[A]$ is decomposable then A is equivalent to a matrix B of the form (3) with $s > 1$ and the $B_{i,i}$ indecomposable. Obviously, $\mathcal{E}(B)$ contains all the matrices of the form

$$\begin{pmatrix} c_1 I_{n_1} & & 0 \\ & \ddots & \\ 0 & & c_s I_{n_s} \end{pmatrix} \tag{4}$$

where n_i denotes the size of the block $B_{i,i}$, I_{n_i} the identity matrix of size n_i and where the c_i's are arbitrary constants. By taking $c_1 = 0$, one sees that $\mathcal{E}(B)$, and then $\mathcal{E}(A)$, is not a division ring. □

Remark 3.1. If a system $[A]$ is irreducible then, by Theorem 3.1, $\mathcal{E}(A)$ is a division ring . The converse is not true. To see this, let $K = \mathbb{C}(x)$, ϕ the automorphism of K defined by $\phi x = x - 1$, $\delta = 1 - \phi$ and $A = \begin{pmatrix} 2 & 0 \\ 1 & x \end{pmatrix}$. One can show, by a direct calculation, that

$$\mathcal{E}(A) = \left\{ \begin{pmatrix} c & 0 \\ 0 & c \end{pmatrix} \text{ where } c \in \mathbb{C} \right\}.$$

However, if $[A]$ is a completely reducible equation which is reducible then its eigenring $\mathcal{E}(A)$ is not a division ring. This follows from the fact that an equation which is at the same time reducible and completely reducible is necessarily decomposable, so by Proposition 3.1 its eigenring is not a division ring.

We have seen, in the proof of Proposition 3.1, that the eigenring of a decomposable equation $[A]$ contains matrices that are similar to the matrices of the form (4). Hence, the eigenring of a decomposable equation contains always a matrix which has at least two distinct eigenvalues. The next thorem shows that the converse is true: if eigenring of $[A]$ contains a matrix T which has at least two distinct eigenvalues then $[A]$ is decomposable; furthermore, it can be decomposed by finding a matrix $P \in GL(n, K)$ such that $P^{-1}TP$ have a block-diagonal form $J_1 \oplus \cdots \oplus J_s$ where the J_i are constant matrices with disjoint spectra.

For the sake of clarity we add the technical assumption that the field of constants $\mathcal{C}$ is an algebraically closed field.

Theorem 3.2. *Assume that the field of constants $\mathcal{C}$ is algebraically closed. Let $[A]$ be a matrix pseudo-linear equation whose eigenring contains an element T which has $s \geq 2$ distinct eigenvalues $\lambda_1, \ldots, \lambda_s \in \mathcal{C}$. Let $P \in GL(n, K)$ such that the matrix J defined by $J = P^{-1}TP$ be of the form*

$$J_1 \oplus \cdots \oplus J_s$$

where J_i is an $n_i \times n_i$ matrix of the form

$$\begin{pmatrix} \lambda_i & \varepsilon & & 0 \\ 0 & \ddots & & \\ & & \ddots & \varepsilon \\ 0 & & 0 & \lambda_i \end{pmatrix} \quad \text{with } \varepsilon = 0 \text{ or } 1.$$

Then the matrix $B = P^{-1}(A\phi P + \delta P)$ has the form

$$B = B_{1,1} \oplus \ldots \oplus B_{s,s}$$

where $B_{i,i}$ is a square matrix of order n_i.

Proof. In one hand, $J \in \mathcal{E}(B)$, since $T \in \mathcal{E}(A)$, so, J satisfies $JB - B\phi J = \delta J$. On the other hand J is a constant matrix, so $\phi(J) = J$ and $\delta J = 0$ and then the relation $JB - B\phi J = \delta J$ reduces to $JB = BJ$. Now let us decompose B in blocks following the same partition as J, i. e. $B = (B_{i,j})_{1 \leq i,j \leq s}$. Then the equation $BJ = JB$ implies

$$J_i B_{i,j} - B_{i,j} J_j = 0, \text{ for all } 1 \leq i, j \leq s.$$

It then follows that $B_{i,j} = 0$ for $i \neq j$. $\qquad\square$

Remark 3.2. In fact, we have the following more precise result on the structure of the matrix B in in Theorem 3.2. Let the block J_i of J be decomposed as $J_i = J_{i,1} \oplus \cdots \oplus J_{i,m_i}$ where $J_{i,k}$ has the form

$$\begin{pmatrix} \lambda_i & 1 & 0 \\ & \ddots & 1 \\ 0 & & \lambda_i \end{pmatrix}.$$

If $B_{i,i} = (B_{i,i}^{h,k})$ is the partition in blocks of $B_{i,i}$ consistent with the above partition of J_i then each block $(B_{i,i}^{h,k})$ is an upper triangular Toeplitz matrix. This follows from the fact that J_i and $B_{i,i}$ commute and a well known result of linear algebra (see Theorem S2.2 in Ref. 11, pp 345-347).

Maximal Decompsition

Suppose that the field of constants $\mathcal{C}$ is algebraically closed. To factorize a given matrix pseudo-linear equation $[A]$ we proceed as follows. First we determine a basis $\{T_1, \ldots, T_d\}$ of $\mathcal{E}(A)$. Then, we choose a *generic* point $(c_1, \ldots, c_d)$ in $\mathcal{C}^d$, that is a point so that the number s of distinct eigenvalues of $T = c_1 T_1 + \ldots + c_d T_d$ be the maximal possible. Then we determine a nonsingular matrix P such that $J := P^{-1} T P$ be in Jordan canonical form. Finally we compute the matrix $B := P^{-1} A \phi P + P^{-1} \delta P$. Let us note that the diagonal blocks of B will be necessarily indecomposable, since s is maximal.

We repeat that this method allows to compute factorizations of matrix pseudo-linear equations that are both completely reducible and reducible.

We conclude this section with two examples to illustrate the above method.

Example 3.2. Take the matrices A and T of Example 3.1. Consider the matrix

$$P = \begin{bmatrix} -2 & -1 & 2 & 0 \\ 1 & 0 & -1 & 0 \\ -2\,x^{-1} & -x^{-1} & 2\,x^{-1} & x^{-1} \\ 2 & 1 & -1 & 0 \end{bmatrix}.$$

One has

$$
J = P^{-1}TP = \begin{bmatrix} 0 & 0 & 0 & 0 \\ 0 & 0 & 0 & 0 \\ 0 & 0 & -2 & 0 \\ 0 & 0 & 0 & -2 \end{bmatrix}.
$$

Here we have $s = 2$ blocks of dimension 2. Now if we compute the matrix $B := P^{-1}A\phi P + P^{-1}\delta P$ we get

$$
B = \begin{bmatrix} 2\,(x-1)^{-1} & (x-1)^{-1} & 0 & 0 \\ -\frac{-3\,x+4+x^2}{x-1} & -\frac{x^2-3\,x+3}{x-1} & 0 & 0 \\ 0 & 0 & -(x-1)^{-1} & -(x-1)^{-1} \\ 0 & 0 & -\frac{1+2\,x^2-4\,x}{x-1} & -\frac{1+x^2-3\,x}{x-1} \end{bmatrix}.
$$

Example 3.3. This example corresponds to a scalar differential equation taken from Ref. 17 (see also Ref. 8). Let $K = \bar{\mathbb{Q}}(x)$, $\phi = 1_K$ and $\delta = \frac{d}{dx}$. Consider the differential system $[A]$ where A is given by

$$
A = \begin{bmatrix} 0 & -1 & 0 & 0 \\ 0 & 0 & -1 & 0 \\ 0 & 0 & 0 & -1 \\ \dfrac{1}{x^8} & -2\,\dfrac{3\,x^2-1}{x^5} & 2\,\dfrac{x^2-1}{x^4} & 6\,\dfrac{1}{x} \end{bmatrix}
$$

The eigenring of A consists of the matrices T of the form

$$
\begin{bmatrix} 7\,c_2 + c_1 & -2\,c_2x^3 - xc_2 & c_2x^4 & c_2x^5 \\ -\frac{c_2}{x^3} & 4\,c_2 + c_1 & xc_2 & 0 \\ 3\,\frac{c_2}{x^4} & -\frac{c_2}{x^3} & 5\,c_2 + c_1 & xc_2 \\ -\frac{c_2(12\,x^2+1)}{x^7} & 2\,\frac{c_2(3\,x^2+2)}{x^4} & -\frac{c_2(2\,x^2-1)}{x^3} & c_1 \end{bmatrix}
$$

Let us choose $(c_1, c_2) = (0, 1)$

$$
T = \begin{bmatrix} 7 & -2\,x^3 - x & x^4 & x^5 \\ -x^{-3} & 4 & x & 0 \\ 3\,x^{-4} & -x^{-3} & 5 & x \\ -\frac{12\,x^2+1}{x^7} & 2\,\frac{3\,x^2+2}{x^4} & -\frac{2\,x^2-1}{x^3} & 0 \end{bmatrix}
$$

We compute the Jordan form of T

$$J = P^{-1}TP = \begin{bmatrix} 4+\sqrt{2} & 0 & 0 & 0 \\ 0 & 4+\sqrt{2} & 0 & 0 \\ 0 & 0 & -\sqrt{2}+4 & 0 \\ 0 & 0 & 0 & -\sqrt{2}+4 \end{bmatrix}$$

where

$$P = \begin{pmatrix} P_{1,1} & P_{1,2} \\ P_{2,1} & P_{2,2} \end{pmatrix}$$

with

$$P_{1,1} = \begin{bmatrix} \left(1/2+\sqrt{2}\right)x^4 + 3/4\sqrt{2} + 1/2\left(1/2+\sqrt{2}\right)x^4 \\ -1/4\frac{\sqrt{2}}{x^3} \qquad\qquad 0 \end{bmatrix}$$

$$P_{1,2} = \begin{bmatrix} \left(-\sqrt{2}+1/2\right)x^4 - 3/4\sqrt{2} + 1/2\left(-\sqrt{2}+1/2\right)x^4 \\ 1/4\frac{\sqrt{2}}{x^3} \qquad\qquad 0 \end{bmatrix}$$

$$P_{2,1} = \begin{bmatrix} \sqrt{2}+1/2+3/4\frac{\sqrt{2}}{x^4} & 1/2+\sqrt{2} \\ -7/2\frac{\sqrt{2}}{x}-3\frac{\sqrt{2}}{x^5}-1/4\frac{\sqrt{2}}{x^7} & -7/2\frac{\sqrt{2}}{x} \end{bmatrix}$$

$$P_{2,2} = \begin{bmatrix} -\sqrt{2}+1/2-3/4\frac{\sqrt{2}}{x^4} & -\sqrt{2}+1/2 \\ 7/2\frac{\sqrt{2}}{x}+3\frac{\sqrt{2}}{x^5}+1/4\frac{\sqrt{2}}{x^7} & 7/2\frac{\sqrt{2}}{x} \end{bmatrix}$$

Now if we compute the matrix $B = P^{-1}(AP+\delta P)$ then we get, as expected, a block-diagonal matrix $B = B_{1,1} \oplus B_{2,2}$. Here the diagonal blocks are respectively

$$\begin{bmatrix} \left(4+\sqrt{2}\right)x^3 & \left(4+\sqrt{2}\right)x^3 \\ \left(-4-\sqrt{2}\right)x^3 + \frac{1-\sqrt{2}}{x} + \frac{4-\sqrt{2}}{14\,x^7} & \left(-4-\sqrt{2}\right)x^3 + \frac{1-\sqrt{2}}{x} \end{bmatrix}$$

$$\begin{bmatrix} x^3\left(4-\sqrt{2}\right) & x^3\left(4-\sqrt{2}\right) \\ \left(-4+\sqrt{2}\right)x^3 + \frac{1+\sqrt{2}}{x} + \frac{4+\sqrt{2}}{14\,x^7} & \left(-4+\sqrt{2}\right)x^3 + \frac{1+\sqrt{2}}{x} \end{bmatrix}.$$

4. Spaces of homomorphisms and factorization

Let $[A]$ and $[B]$ be two matrix pseudo-linear equations of dimension m and n respectively. We define $\mathcal{H}(A, B)$ as the set of $n \times m$ matrices T in K satisfying the equation

$$TA - B\phi(T) = \delta T.$$

This equation can be viewed as a system of nm first order pseudo-linear equations over (K, ϕ, δ). As a consequence, $\mathcal{H}(A, B)$ is a $\mathcal{C}-$vector space of finite dimension $(\leq mn)$. Note that when $A = B$, $\mathcal{H}(A, B) = \mathcal{E}(A)$. A simple computation shows that if y satisfies $[A]$ then Ty satisfies $[B]$, for all $T \in \mathcal{H}(A, B)$. Thus, the elements of $\mathcal{H}(A, B)$ transform a solution of the matrix pseudo-linear equation $[A]$ into a solution of $[B]$. So, it is natural to expect that the properties of $\mathcal{H}(A, B)$ will be useful in studying reducibility of the equations $[A]$ and $[B]$.

Proposition 4.1. *Let $[A]$ and $[B]$ be two matrix pseudo-linear equations of dimension m and n with $m < n$. Suppose that there exists $T \in \mathcal{H}(A, B)$ such that $\operatorname{rank} T = m$. Then $[B]$ is reducible and one can construct a transformation $P \in GL(n, K)$ such that the matrix $P^{-1}(B\phi P + \delta P)$ have the form*

$$\begin{pmatrix} C_{1,1} & 0 \\ C_{2,1} & A \end{pmatrix},$$

where $C_{1,1}$ is an $(n - m) \times (n - m)$ matrix.

Proof. One can compute an $n \times (n - m)$ matrix S in K such that the $n \times n$ matrix defined by $P = (S \ \ T)$ be invertible. Let $C = P^{-1}(B\phi P + \delta P)$. Decompose C as

$$C = \begin{pmatrix} C_{1,1} & C_{1,2} \\ C_{2,1} & C_{2,2} \end{pmatrix},$$

where $C_{1,1}$ is a square matrix of size $n - m$. We shall show that $C_{1,2} = 0$ and $C_{2,2} = A$.

First, observe that the relation $PC - B\phi P = \delta P$, implies

$$P \begin{pmatrix} C_{1,2} \\ C_{2,2} \end{pmatrix} = B\phi T - \delta T.$$

Since $T \in \mathcal{H}(A, B)$, one has $B\phi T - \delta T = TA$. This can be rewritten as

$$B\phi T - \delta T = TA = P \begin{pmatrix} 0 \\ A \end{pmatrix}.$$

Thus,

$$P\begin{pmatrix} C_{1,2} \\ C_{2,2} \end{pmatrix} = P\begin{pmatrix} 0 \\ A \end{pmatrix}.$$

It then follows that $C_{1,2} = 0$ and $C_{2,2} = A$, since P is invertible.　　□

Example 4.1. Let $K = \bar{\mathbb{Q}}(x)$, ϕ the automorphism of K defined by $\phi x = x - 1$ and $\delta = 1 - \phi$ and consider the matrices

$$A = \begin{bmatrix} -x + 2 & -x \\ -(x-1)^{-1} & 0 \end{bmatrix},$$

$$B = \begin{bmatrix} (x-1)^{-1} & -1 & 1 & \frac{x+1}{x-1} \\ 0 & 1 & -1 & -\frac{x}{x-1} \\ \frac{x-2}{x} & -2\,x^{-1} & 3-x & -\frac{-3\,x+x^2+1}{(x-1)x} \\ 0 & x-1 & -1 & -\frac{2+x^2-2\,x}{x-1} \end{bmatrix}.$$

Here $m = 2$ and $n = 4$. $\mathcal{H}(A, B)$ consists in the matrices of the form

$$\begin{bmatrix} 0 & c_1 \\ 0 & -c_1 \\ -\frac{c_1}{x} & \frac{c_1}{x} \\ 0 & -c_1 \end{bmatrix}$$

where c_1 is an arbitrary constant. Let T be the element of $\mathcal{H}(A, B)$ obtained by taking $c_1 = 1$. Then T has rank 2. We complete T in order to get an invertible matrix P of size 4 :

$$P := \begin{bmatrix} 1 & 0 & 0 & 1 \\ 0 & 1 & 0 & -1 \\ 0 & 0 & -\frac{1}{x} & \frac{1}{x} \\ 0 & 0 & 0 & -1 \end{bmatrix}$$

If we compute the matrix $C := P^{-1}(B\phi P + \delta P)$, we get as expected

$$C = \begin{bmatrix} (x-1)^{-1} & x-2 & 0 & 0 \\ 0 & 2-x & 0 & 0 \\ 2-x & -x+3 & 2-x & -x \\ 0 & -x+1 & -(x-1)^{-1} & 0 \end{bmatrix}.$$

Let us note that the equation $[B]$, of the above example, is reducible but it cannot be reduced using the method exposed in section 3. Indeed, the computation of the eigenring of B shows that $\dim(\mathcal{E}(B)) = 1$.

The following proposition is a generalization of Theorem 3.1.

Proposition 4.2. *Let $[A]$ and $[B]$ be two pseudo-linear equations of dimension m and n with $m \leq n$. Suppose that there exists $T \in \mathcal{H}(A, B)$ such that $0 < rankT = r < m$. Then the systems $[A]$ and $[B]$ are reducible. Moreover, one can construct two matrices $Q \in GL(m, K)$ and $P \in GL(n, K)$ such that*

$$Q^{-1}A\phi Q + Q^{-1}\delta Q = \begin{pmatrix} D_{1,1} & 0 \\ D_{2,1} & D_{2,2} \end{pmatrix}$$

where $D_{1,1}$ is an $r \times r$ matrix, and

$$P^{-1}B\phi P + P^{-1}\delta P = \begin{pmatrix} C_{1,1} & 0 \\ C_{2,1} & C_{2,2} \end{pmatrix},$$

where $C_{2,2} = D_{1,1}$.

Proof. Since $0 < r = rank(T) < m$, one can compute a matrix $Q \in GL(m, K)$ such that

$$S := TQ = \begin{pmatrix} S_{1,1} & 0 \\ S_{2,1} & 0 \end{pmatrix},$$

where $S_{1,1}$ is an $r \times r$ matrix. Let

$$D := Q^{-1}(A\phi Q + \delta Q) = \begin{pmatrix} D_{1,1} & D_{1,2} \\ D_{2,1} & D_{2,2} \end{pmatrix},$$

$D_{1,1}$ being a square matrix of size r. Then one has $S = TQ \in \mathcal{H}(D, B)$. Using the relation $SD - B\phi(S) = \delta S$, one sees that

$$\begin{pmatrix} S_{1,1} \\ S_{2,1} \end{pmatrix} D_{1,2} = 0$$

and

$$\begin{pmatrix} S_{1,1} \\ S_{2,1} \end{pmatrix} D_{1,1} - B \begin{pmatrix} \phi S_{1,1} \\ \phi S_{2,1} \end{pmatrix} = \begin{pmatrix} \delta S_{1,1} \\ \delta S_{2,1} \end{pmatrix}.$$

The first equation implies that $D_{1,2} = 0$, since the matrix $U := \begin{pmatrix} S_{1,1} \\ S_{2,1} \end{pmatrix}$ has full rank r and the second equation implies that $U \in \mathcal{H}(D_{1,1}, B)$. One can apply Proposition 4.1 to the systems $[D_{1,1}]$ and $[B]$ to achieve the proof.$\square$

Corollary 4.1. *Assume that the field of constants C is algebraically closed. If $[A]$ and $[B]$ are equivalent and $\dim \mathcal{H}(A, B) > 1$ then $[A]$ and $[B]$ are both reducible.*

Proof. Suppose that $A \sim B$ then, there exists a nonsingular matrix $P \in \mathcal{H}(A, B)$. Let $Q \in \mathcal{H}(A, B)$ such that the matrices Q and P be linearly independent over C (such a matrix exists since $\dim \mathcal{H}(A, B) > 1$). Consider the matrix $T = Q - \lambda P$ where λ is any eigenvalue of $P^{-1}Q$. Note that $\lambda \in C$ for $P^{-1}Q \in \mathcal{E}(A)$. So, $T \in \mathcal{H}(A, B)$. One has $\det T = \det P \det (P^{-1}Q - \lambda I) = 0$. Thus, $\mathcal{H}(A, B)$ contains a non-zero matrix T with rank $< n$ (n being the order of the matrix A). Hence $[A]$ and $[B]$ are reducible according to the above proposition. $\square$

As a consequence we have the following result whose proof is immediate.

Corollary 4.2. *Assume that the field of constants C is algebraically closed. Let $[A]$ and $[B]$ be two pseudo-linear equations of dimension m and n with $m \le n$. Suppose that the equation $[B]$ is irreducible. Then $\mathcal{H}(A, B)$ has dimension 1 or 0, depending on whether $[A]$ and $[B]$ are equivalent or not.*

Appendix A. $K[X; \phi, \delta]$−modules and matrix pseudo-linear equations

In this section we shall see that a matrix pseudo-linear equation over (K, ϕ, δ) can be thought of as a left module over the Ore polynomial ring $K[X; \phi, \delta]$.[13] This classical approach has the advantage of enabling one to apply directly the general theorems on modules (like the Jordan-Hölder theorem, Schur's lemma, the Krull-Schmidt theorem[14]) to matrix pseudo-linear equations. This allows a better understanding of the problems arising in the study of matrix pseudo-linear equations. However, the results we shall recall in this section are not required for the rest of the paper.

Appendix A.1.   *Pseudo-linear operators*

Let V be a vector space of finite dimension n over K. A *pseudo-linear operator* on V with respect to (K, ϕ, δ) is a C−linear map $L : V \longrightarrow V$ satisfying $L(av) = \phi a \, Lv + \delta a \, v$ for all $a \in K$, $v \in V$.
Let L be a pseudo-linear operator on V and $(e_1, \ldots, e_n)$ a basis of V. The matrix of L w.r.t. this basis is defined as the $n \times n$ matrix $A = (a_{ij})$ where

$Le_j = \sum\limits_{i=1}^{n} a_{ij}e_i$, for $j = 1, \ldots, n$. If $v = \sum v_i e_i$ is any vector in V and $Lv = \sum w_i e_i$, then

$$\begin{pmatrix} w_1 \\ \vdots \\ w_n \end{pmatrix} = \begin{pmatrix} \delta\,v_1 \\ \vdots \\ \delta\,v_n \end{pmatrix} + A \begin{pmatrix} \phi\,v_1 \\ \vdots \\ \phi\,v_n \end{pmatrix}. \tag{A.1}$$

Thus L is completely determined by its matrix in a fixed basis. Conversely, if A is an $n \times n$ matrix in K the map $v \mapsto Lv$ defined by (A.1) is a pseudo-linear operator of V. Thus with any matrix pseudo-linear equation one can associate a pseudo-linear operator and conversely.

Let $(f_1, \ldots, f_n)$ be a second basis of V related to $(e_1, \ldots, e_n)$ by

$$(f_1, \ldots, f_n) = (e_1, \ldots, e_n)P$$

where $P \in GL(n, K)$. If A and B are respectively the matrices of L relative to the bases $(e_1, \ldots, e_n)$ and $(f_1, \ldots, f_n)$ then

$$B = P^{-1}A\phi P + P^{-1}\delta P.$$

Let us point out that pseudo-linear operators has been considered and studied by Jacobson in Ref. 13. He introduce the notion of *elementary divisors* of pseudo-linear operators and shows that as in the theory of linear operators one can give criteria for similarity, reducibility, decomposability, and complete reducibility in terms of elementary divisors. These results are not used in the present paper.

Appendix A.2. *Similarity, reducibility, decomposability and complete reducibility*

Let $K[X; \phi, \delta]$ denote the ring of non-commutative polynomials in the indeterminate X with coefficients taken on the left in K with the usual addition and in which multiplication is defined by $Xa = (\phi a)X + \delta a$ for all $a \in K$. Let V be a vector space of finite dimension n over K and L a pseudo-linear operator on V with respect to (K, ϕ, δ). One can define $Xv = L(v)$ for all $v \in V$. This makes V into a left $K[X; \phi, \delta]$−module. We shall denote this module (V, L). Conversely, if V is a left $K[X; \phi, \delta]$−module and a finite dimensional K-vector space, then by $L(v) = Xv$ for all $v \in V$ a pseudo-linear operator is defined.

Thus with any matrix pseudo-linear equation one can associate a $K[X; \phi, \delta]$−module and conversely.

Two pseudo-linear operators on V, L and M, are said to be *similar* if (V, L) is isomorphic, as a $K[X; \phi, \delta]$−module, to (V, M). A simple computation shows that L and M are similar if, and only if, their matrices, relative to a given basis of V, are equivalent.

Let L be a pseudo-linear operator L on V. A subspace $W \subset V$ is called *invariant* if $LW \subset W$. One can see easily that $W \subset V$ is invariant if, and only if, W is a submodule of (V, L).

The operator L is called *reducible* if V is reducible as a $K[X; \phi, \delta]$−module, i.e. if there exists a submodule W of V such that $0 \neq W \neq V$. Otherwise, L is called *irreducible*.

The operator L is called *decomposable* if V is decomposable as a $K[X; \phi, \delta]$−module, i.e. if $V = W_1 \oplus W_2$ where $W_i \neq 0$. Otherwise, L is called *indecomposable*.

The operator L is called *completely reducible* if V is completely reducible as a $K[X; \phi, \delta]$−module, i.e. if V is a direct sum of irreducible submodules. In matrix terms: if A denotes the matrix of L w.r.t. a $K-$ basis of V, then L is reducible, decomposable or completely reducible if, and only if, the matrix equation $[A]$ is respectively reducible, decomposable or completely reducible.

Now the general theorems on modules translated in matrix terms read:

Jordan-Hölder. Any matrix pseudo-linear equation $[A]$ is equivalent to a matrix equation $[B]$ having the form

$$\begin{pmatrix} B_{1,1} & 0 & & 0 \\ B_{2,1} & B_{2,2} & & \\ \vdots & & \ddots & 0 \\ B_{k,1} & \cdots & & B_{k,k} \end{pmatrix}$$

where for each $1 \leq i \leq k$, the matrix equation $[B_{i,i}]$ is irreducible. Moreover, the the the set of integers $\{k, n_1, \ldots, n_k\}$, where n_i denotes the size of $B_{i,i}$ is uniquely determined by A.

Krull-Schmidt. Any matrix pseudo-linear equation $[A]$ is equivalent to a matrix equation $[B]$ having the form

$$\begin{pmatrix} B_{1,1} & & 0 \\ & \ddots & \\ 0 & & B_{s,s} \end{pmatrix}$$

where for each $1 \leq i \leq k$, the matrix equation $[B_{i,i}]$ is indecomposable. Moreover, the the set of integers $\{s, n_1, \ldots, n_s\}$, where n_i denotes the size of $B_{i,i}$, is uniquely determined by A.

Krull-Schmidt theorem for completely reducible modules. If A is equivalent to matrix of of the form (3) where the $[B_{i,i}]$ are irreducible, then the set of integers $\{s, n_1, \ldots, n_s\}$, where n_i denotes the size of $B_{i,i}$, is uniquely determined by A.

Appendix A.3. *The ring of endomorphisms of a pseudo-linear operator*

Let V be a vector space of finite dimension n over K and L a pseudo-linear operator on V with respect to (K, ϕ, δ).

The ring of endomorphisms of L is by definition the ring of endomorphisms of the module (V, L). Let $\mathcal{E}nd(L)$ denote this ring. It is not difficult to see that a map $u : V \longrightarrow V$ belongs to $\mathcal{E}nd(L)$ if, and only if, u is a $K-$linear map of V satisfying $u \circ L = L \circ u$. This condition in term of the matrices A, T of L, u, w.r.t. a $K-$ basis of V, is

$$\delta T = TA - A\phi(T).$$

It follows that $\mathcal{E}nd(L)$ is isomorphic to $\mathcal{E}(A)$.

As a consequence we get that the eigenrins of two similar operators are isomorphic.

Using Schur's Lemma one can prove that if L is irreducible then $\mathcal{E}nd(L)$ is a division ring (compare with our Theorem 3.1).

For more details on ring of endomorphisms see Ref. 13 (pp 502-503) where a complete description of the ring of endomorphisms of an irreducible operator is given.

References

1. S.A. Abramov. Rational solutions of linear difference and q-difference equations with polynomial coefficients. In *Proceedings of ISSAC'95*, pages 303–308. ACM Press, 1995.
2. S.A. Abramov. EG-eliminations. *Journal of Difference Equations and Applications*, 5:393–433, 1999.
3. S.A. Abramov and M.A. Barkatou. Rational Solutions of First Order Linear Difference Systems. In *Proceedings of ISSAC'98*, pages 124–131. ACM Press, 1998.
4. M. Barkatou, T. Cluzeau, and JA. Wei. *Factoring Partial Differential Systems in Positive Characterestic*, pages 213–238. Birkhauser, 2005.
5. M.A. Barkatou. On rational solutions of systems of linear differential equations. *J. Symbolic Computation*, 28(4 and 5):547–567, 1999.
6. M.A. Barkatou. Rational solutions of matrix difference equation. problem of equivalence and factorisation. In *Proceedings of ISSAC'99*, pages 167–175. ACM Press, 1999.

7. M.A. Barkatou. On the reduction of matrix pseudo-linear equations. Technical Report RR 1040, IMAG Grenoble, France, may 2001.

8. M.A. Barkatou and E. Pflügel. On the Equivalence Problem of Linear Differential Systems and its Application for Factoring Completely Reducible Systems. In *Proceedings of ISSAC'98*, pages 268–275. ACM Press, 1998.

9. M.A. Barkatou and E. Pflügel. An algorithm computing the regular Formal Solutions of a system of linear Differential Equations. *Journal of Symbolic Computation*, 28(4 and 5):569–587, 1999.

10. T. Cluzeau. Factorization of differential systems in characteristic *p*. In *Proceedings of ISSAC'2003*, pages 58–65. ACM Press, 2003.

11. I. Gohberg, P. Lancaster, and L. Rodman. *Matrix polynomials*. Academic Press, New York; London; Paris, 1982.

12. D. Grigoriev. Complexity of Irreducibility Testing for a System of Linear Ordinary Differential Equations. In *Proceedings of ISSAC'90*. ACM Press, 1990.

13. N. Jacobson. Pseudo-linear transformations. *Annals of Mathematics*, 38(2):484–507, 1937.

14. N. Jacobson. *Basic Algebra II*. W. H. Freeman and Company, San Fransisco, 1980.

15. M. Singer. *Direct and Inverse Problems in Differential Galois Theory*, pages 527–554. American Mathematical Society, 1999.

16. M.F. Singer. Testing Reducibility of Linear Differential Operators: A Group Theoretic Perspective. *J. of Appl. Alg. in Eng. Comm. and Comp.*, 7(2):77–104, 1996.

17. M. van Hoeij. Rational Solutions of the Mixed Differential Equation and its Application to Factorization of Differential Operators. In *Proceedings of ISSAC'96*. ACM Press, 1996.

18. M. van Hoeij. Rational solutions of linear difference equations. In *Proceedings of ISSAC'98*, pages 120–123. ACM Press, 1998.

MODULAR COMPUTATION FOR MATRICES OF ORE POLYNOMIALS

HOWARD CHENG

Department of Mathematics and Computer Science
University of Lethbridge
Lethbridge, Canada
E-mail: howard.cheng@uleth.ca

GEORGE LABAHN

David R. Cheriton School of Computer Science
University of Waterloo
Waterloo, Canada
E-mail: glabahn@uwaterloo.ca

We give a modular algorithm to perform row reduction of a matrix of Ore polynomials with coefficients in $\mathbb{Z}[t]$. Both the transformation matrix and the transformed matrix are computed. The algorithm can be used for finding the rank and left nullspace of such matrices. In the special case of shift polynomials, we obtain algorithms for computing a weak Popov form and for computing a greatest common right divisor (GCRD) and a least common left multiple (LCLM) of matrices of shift polynomials. Our algorithms improve on existing fraction-free algorithms and can be viewed as generalizations of the work of Li and Nemes on GCRDs and LCLMs of Ore polynomials. We define lucky homomorphisms, determine the appropriate normalization, as well as bound the number of homomorphic images required. Our algorithm is output-sensitive, such that the number of homomorphic images required depends on the size of the output. Furthermore, there is no need to verify the result by trial division or multiplication. When our algorithm is used to compute a GCRD and a LCLM of shift polynomials, we obtain a new output-sensitive modular algorithm.

1. Introduction

Ore polynomials provide a general setting for describing linear differential, difference and q-difference operators.[15] Systems of equations defined by these operators can be represented by matrices of Ore polynomials. In this paper we look at the problem of transforming such matrices into "simpler" ones using only certain row operations. Examples of such transformations include conversion to special forms, such as row-reduced and weak Popov normal forms.[3,6,10,14]

Performing row reductions on a matrix of Ore polynomials to these simpler forms allows one to determine its rank and left nullspace, giving the minimum number of equations needed to represent the system of equations.[2,6] If the transformation is invertible, the normal form in fact gives the matrix representing an equivalent system with a minimum number of equations. When the leading coefficient is triangular (as in the weak Popov form), the normal form allows one to rewrite high-order operators (e.g. derivatives) in terms of lower ones (Example 2.1). These transformations can also be applied to the computation of greatest common right divisors (GCRDs) and least common left multiples (LCLMs),[3,6,11–13] which represents the intersection and the union of the solution spaces of systems of equations.

The general problem of row reduction was discussed in Refs. 3 and 6. The problem of defining and computing Popov forms over non-commutative valuation domains such as rings of Ore polynomials was considered in Ref. 9, but efficient computation of Popov forms is not considered. In practice, row reductions can introduce significant coefficient growth which must be controlled. This is important in the case of Ore polynomials as coefficient growth is introduced in two ways—from multiplying on the left by powers of the indeterminate and from elimination by cross-multiplication.

Fraction-free algorithms were given in Refs. 3 and 6 to compute the rank and a left nullspace of matrices of Ore polynomials. When the matrix entries are shift polynomials, we obtained fraction-free algorithms for computing row-reduced and weak Popov forms, and for computing a greatest common right divisor (GCRD) or a least common left multiple (LCLM) of matrices of shift polynomials. It was shown that the fraction-free algorithms can be viewed as a generalization of the subresultant algorithm of Li[11,12] to the case of matrices. Fraction-free methods allow us to control the growth of intermediate results at a reasonable cost, leading to polynomial time algorithms.

Modular computation is generally faster than the corresponding fraction-free computation in a number of problems.[8,17] In this paper we are interested in modular algorithms for row reduction of matrices of Ore polynomials, as well as computing the associated transformation matrices. There are traditionally three major issues that must be addressed: the problem of "unlucky" homomorphisms, the number of images required for the reconstruction of the result, and the normalization of the result to compute consistent images under different homomorphisms.

In the case of polynomial matrices, these issues are overcome by formulating the row reduction problem as a linear algebra problem. We obtained a modular algorithm which has a lower complexity than the fraction-free counterpart.[7] The algorithm is output-sensitive, so that the number of homomorphic images required depends on the size of the output. Furthermore, there is no need to verify the result by trial division or multiplication. We note that in many modular algorithms, this verification step can be a significant bottleneck. Experimental results showed that when the matrix entries are the commutative polynomials with coefficients in $\mathbb{Z}[t]$, the output-sensitive modular algorithm is significantly superior to the fraction-free algorithm. For matrices of Ore polynomials, the obvious modular algorithm consists of two parts. First, the problem in $\mathbb{Z}[t][Z; \sigma, \delta]$ are reduced to problems in $\mathbb{Z}_p[t][Z; \sigma, \delta]$. Next, evaluation maps $t \leftarrow \alpha$ are applied to reduce the problems to ones whose coefficients are in $\mathbb{Z}_p$. However, such evaluations are generally not Ore ring homomorphisms. As a result, it is not possible to apply the same technique used in the polynomial matrix case.

For the problem of computing GCRDs and LCLMs of Ore polynomials, Li mapped the problem into a linear algebra problem over $\mathbb{Z}_p$ by applying the evaluation map to the entries of the Sylvester matrix of the input polynomials.[13] In the case of Ore polynomial matrices, however, the dimensions and the configuration of the final coefficient matrix (the striped Krylov matrix[3,6]) are not known a priori. Thus, the approach of Li[13] cannot be applied directly in our case.

The purpose of this paper is to overcome these difficulties for matrices of Ore polynomials. We show how these issues are resolved by studying the linear algebra formulation of the problem. We also extend the approach of Cabay[5] to obtain output-sensitive algorithms which do not require trial division or multiplication to verify the results. The complexity of the new modular algorithm improves on the complexity of the fraction-free algorithm. Furthermore, the algorithm can be significantly faster when the size of output is small. We also obtain a new output-sensitive modular algorithm for computing GCRDs and LCLMs of shift polynomials. Experimental results confirm the performance of the modular algorithm as predicted by the complexity analysis.

The paper is organized as follows. In Sec. 2, we review the relevant definitions of matrices of Ore polynomials as well as the fraction-free elimination algorithm of Refs. 3 and 6. In Sec. 3, we give a linear algebra formulation of the problem and illustrate the difficulties in obtaining a modular algorithm. We then discuss the reduction $\mathbb{Z}[t][Z; \sigma, \delta] \rightarrow \mathbb{Z}_p[t][Z; \sigma, \delta]$. The techniques are then extended to the reduction to linear algebra problems in $\mathbb{Z}_p$ in the next section. In Sec. 6 we study the complexity of the algorithms presented. Implementation considerations and experimental results are discussed in Sec. 7. Concluding remarks are discussed in the closing section.

2. Preliminaries

2.1. *Notation*

For any matrix A, we denote its elements by $A_{i,j}$. For any sets of row and column indices I and J, we denote by $A_{I,J}$ the submatrix of A consisting of the rows and columns indexed by I and J. For convenience, we use $*$ for I or J to denote the sets of all rows and columns.

For any vector of integers (also called *multi-index*) $\vec{\omega} = (\omega_1, \ldots, \omega_p)$, we denote by $|\vec{\omega}| = \sum_{i=1}^{p} \omega_i$. The function $\max(\cdot, \cdot)$ gives the vector whose components are the maximum of the corresponding components of its input vectors. We say that $\vec{v} \leq_{\text{lex}} \vec{w}$ if $\vec{v} = \vec{w}$ or if the leftmost nonzero entry in $\vec{v} - \vec{w}$ is negative. The vector $\vec{e}_i$ denotes the i-th unit vector (of the appropriate dimension) such that $(e_i)_j = \delta_{ij}$; we also have $\vec{e} = (1, \ldots, 1)$ (of the appropriate dimension). We denote by $\mathbf{I}_m$ the $m \times m$ identity matrix, and by $Z^{\vec{\omega}}$ the diagonal matrix having Z^{ω_i} on the diagonal.

2.2. *Definitions*

We first give some definitions on Ore polynomial matrices. These definitions are similar to those given in our previous work.[3,6,7]

We denote by $\mathbb{Z}$ the ring of integers, $\mathbb{Q}$ the field of rational numbers, and $\mathbb{Z}_p$ the finite field of p elements where p is prime.

Let k be any field and let $\sigma : k \to k$ be an injective endomorphism of k. Then, $\delta : k \to k$ is a *derivation* with respect to σ is an endomorphism of the additive group of k satisfying

$$\delta(rs) = \sigma(r)\delta(s) + \delta(r)s$$

for all r, $s \in k$. In this paper, we will examine Ore polynomial rings with coefficients in $\mathbb{Z}[t]$. That is, the ring $\mathbb{Z}[t][Z; \sigma, \delta]$ with σ an automorphism, δ a derivation and with the multiplication rule

$$Z \cdot a = \sigma(a)Z + \delta(a)$$

for all $a \in \mathbb{Z}[t]$. When $\delta = 0$, we call the polynomials *shift polynomials*. For brevity, we will use $\mathbb{Z}[t][Z]$ when the specific choices of σ and δ are not important.

Let $\mathbb{Z}[t][Z]^{m \times n}$ be the ring of $m \times n$ Ore polynomial matrices over $\mathbb{Z}[t]$. We shall adapt the following conventions for the remainder of this paper. Let $\mathbf{F}(Z) \in \mathbb{Z}[t][Z]^{m \times n}$, $N = \deg \mathbf{F}(Z)$, and write

$$\mathbf{F}(Z) = \sum_{j=0}^{N} F^{(j)} Z^j, \text{ with } F^{(j)} \in \mathbb{Z}[t]^{m \times n}.$$

We also write $c_j\left(\mathbf{F}(Z)\right) = F^{(j)}$ to denote the j-th coefficient matrix. An Ore polynomial matrix $\mathbf{F}(Z)$ is said to have *row degree* $\vec{\nu} = \text{rdeg}\,\mathbf{F}(Z)$ if the i-th row has degree ν_i. The *leading row coefficient* of $\mathbf{F}(Z)$, denoted $\text{LC}_{\text{row}}\left(\mathbf{F}(Z)\right)$, is the matrix whose entries are the coefficients of Z^N of the corresponding elements of $Z^{N \cdot \vec{e} - \vec{\nu}} \cdot \mathbf{F}(Z)$. An Ore polynomial matrix $\mathbf{F}(Z)$ is *row-reduced* if $\text{LC}_{\text{row}}\left(\mathbf{F}(Z)\right)$ has maximal row rank. We also recall that the *rank* of $\mathbf{F}(Z)$ is the maximum number of $\mathbb{Q}[t][Z]$-linearly independent rows of $\mathbf{F}(Z)$, and that $\mathbf{U}(Z) \in \mathbb{Z}[t][Z]^{m \times m}$ is *unimodular* if there exists $\mathbf{V}(Z) \in \mathbb{Q}[t][Z]^{m \times m}$ such that $\mathbf{V}(Z) \cdot \mathbf{U}(Z) = \mathbf{U}(Z) \cdot \mathbf{V}(Z) = \mathbf{I}_m$. Some useful properties of matrices of Ore polynomials, such as linear independence and rank, can be found in Ref.[3]

Example 2.1. Consider the differential algebraic system

$$
\begin{array}{llll}
y_1''(t) + (t+2)y_1(t) + & y_2''(t) + y_2(t) + & y_3'(t) + y_3(t) = 0 \\
y_1''(t) + y_1'(t) + 3y_1(t) + & y_2^{(3)}(t) + 2y_2'(t) - y_2(t) + & y_3^{(3)}(t) - 2t^2 y_3(t) = 0 \\
y_1'(t) + y_1(t) + & y_2^{(3)}(t) + 2ty_2'(t) - y_2(t) + & y_3^{(4)}(t) = 0.
\end{array}
\tag{1}
$$

Let D denote the differential operator on $\mathbb{Q}(t)$ such that $D \cdot f(t) = \frac{d}{dt} f(t)$. Then the matrix form of Eq. (1) is:

$$
\begin{bmatrix}
D^2 + (t+2) & D^2 + 1 & D + 1 \\
D^2 + D + 3 & D^3 + 2D - 1 & D^3 - 2t^2 \\
D + 1 & D^3 + 2tD + 1 & D^4
\end{bmatrix}
\cdot
\begin{bmatrix}
y_1(t) \\
y_2(t) \\
y_3(t)
\end{bmatrix}
= \mathbf{0}.
\tag{2}
$$

The leading row coefficient (matrix of coefficients of the highest power of the corresponding row) is upper triangular. This allows us to rewrite the highest derivative in each row as a combination of other derivatives. For example, we can eliminate the highest derivatives of $y_2(t)$ as follows:

$$
\begin{aligned}
y_2^{(3)}(t) &= -y_1''(t) - y_1'(t) - 3y_1(t) - 2y_2'(t) + y_2(t) - y_3^{(3)}(t) + 2t^2 y_3(t) \\
&= -((t+2)y_1(t) - y_2''(t) - y_2(t) - y_3'(t) - y_3(t)) - y_1'(t) - 3y_1(t) \\
&\quad - 2y_2'(t) + y_2(t) - y_3^{(3)}(t) + 2t^2 y_3(t) \\
&= -y_1'(t) - (t+5)y_1(t) + y_2''(t) - 2y_2'(t) + 2y_2(t) - y_3^{(3)}(t) + y_3'(t) \\
&\quad + (2t^2 + 1)y_3(t).
\end{aligned}
$$

2.3. The FFreduce Elimination Algorithm

We give a brief description of the FFreduce elimination algorithm[3,6] which forms much of the basis of our work. In this algorithm, we are interested in applying the following elementary row operations to the matrix $\mathbf{F}(Z)$:

(a)   interchange two rows;
(b)   multiply a row by a nonzero element in $\mathbb{Z}[t][Z]$;
(c)   add a polynomial multiple of one row to another.

Formally, we may view a sequence of elementary row operations as a *transformation matrix* $\mathbf{U}(Z) \in \mathbb{Z}[t][Z]^{m \times m}$ with the result of these operations given by $\mathbf{T}(Z) = \mathbf{U}(Z) \cdot \mathbf{F}(Z)$. The application of these row operations do not change the rank of $\mathbf{F}(Z)$.[3,6] If the row multiplier of (b) is restricted to elements of $\mathbb{Z}[t]$ then $\mathbf{U}(Z)$ is unimodular. By applying these operations to eliminate low-order coefficients, one can compute the rank and the left nullspace of $\mathbf{F}(Z)$.

The elimination problem can be formalized as follows. An Ore polynomial vector $\mathbf{P}(Z) \in \mathbb{Z}[t][Z]^{1 \times m}$ is said to have *order* $\vec{\omega}$ with respect to $\mathbf{F}(Z)$[a] if

$$
\mathbf{P}(Z) \cdot \mathbf{F}(Z) = \mathbf{R}(Z) \cdot Z^{\vec{\omega}} \tag{3}
$$

for some *residual* $\mathbf{R}(Z)$. The set of all vectors of a particular order $\vec{\omega}$ forms a $\mathbb{Q}[t][Z]$-module. The FFreduce algorithm computes a basis $\mathbf{M}(Z) \in \mathbb{Z}[t][Z]^{m \times m}$ of row degree $\vec{\mu}$ for this module, called an *order basis*, such that

(1)   every row, $\mathbf{M}(Z)_{i,*}$, has order $\vec{\omega}$ for all $1 \leq i \leq m$;
(2)   the rows of $\mathbf{M}(Z)$ form a basis of the module of all vectors of order $\vec{\omega}$. That is, every $\mathbf{P}(Z) \in \mathbb{Q}[t][Z]^{1 \times m}$ of order $\vec{\omega}$ can be written as $\mathbf{P}(Z) = \mathbf{Q}(Z) \cdot \mathbf{M}(Z)$ for some $\mathbf{Q}(Z) \in \mathbb{Q}[t][Z]^{1 \times m}$;

---

[a]Orders in this paper will be with respect to $\mathbf{F}(Z)$ and it will not be explicitly stated for the remainder of the paper.

(3) the leading column coefficient is normalized. That is, there exists a nonzero $d \in \mathbb{Z}[t]$ such that

$$\mathbf{M}(Z) = d \cdot Z^{\vec{\mu}} + \mathbf{L}(Z)$$

where $\deg \mathbf{L}(Z)_{k,l} \leq \mu_l - 1$.

If $\mathbf{M}(Z)$ is row-reduced, we say that it is a *reduced order basis*. Condition (3) implies that the row degree can be viewed as the number of times each row of $\mathbf{F}(Z)$ has been used as a pivot in the elimination process (see Example 3.2). An order basis of a particular order and degree, if it exists, is unique up to a constant multiple (see Theorem 4.4 in Ref. 3). In the FFreduce algorithm, the order is given as input but the degree of the order basis computed is not known in advance. The final row degree reached depends on the input and a pivoting scheme (also called a *computational path*) to be described later.

Let $\mathbf{R}(Z)$ be the residual corresponding to an order basis $\mathbf{M}(Z)$ of order $\vec{\omega} = \sigma \cdot \vec{e}$, such that

$$\mathbf{M}(Z) \cdot \mathbf{F}(Z) = \mathbf{R}(Z) \cdot Z^{\vec{\omega}} \tag{4}$$

If $\sigma = mN + 1$, then the number of nonzero rows in $\mathbf{R}(Z)$ is $\operatorname{rank}\mathbf{F}(Z)$, and the rows in $\mathbf{M}(Z)$ corresponding to the zero rows in the residual gives a basis of the left nullspace of $\mathbf{F}(Z)$.[3,6] The order basis computed has row degree $\vec{\mu}$ such that $\mu_i \leq (mN + 1)n$. For the remainder of this paper, we only consider the order basis problem in Eq. (4) with $\vec{\omega} = \sigma \cdot \vec{e}$.

If $\mathbf{F}(Z)$ is a matrix of shift polynomials, we have $|\vec{\mu}| \leq \sigma \min(m, n)$, $\mu_i \leq \sigma$, and the trailing coefficient of $\mathbf{R}(\mathbf{Z})$ has rank $\mathbf{F}(Z)$ nonzero rows. In fact, the reduction can be terminated as soon as there are "enough" zero rows in $\mathbf{R}(Z)$.[2,3,6] In this case, one may perform row reduction on $\mathbf{F}(Z) \cdot Z^{-N}$ using Z^{-1} as the indeterminate. Reversing the coefficients of $\mathbf{M}(Z^{-1})$ and $\mathbf{R}(Z^{-1})$, we get

$$\mathbf{U}(Z) \cdot \mathbf{A}(Z) = \mathbf{T}(Z) \tag{5}$$

with $\mathbf{U}(Z)$ unimodular and $\mathbf{T}(Z)$ in row-reduced form. We can also choose the pivot rows used in the last n steps of the algorithm to construct $\hat{\mathbf{M}}(Z^{-1})$ and $\hat{\mathbf{R}}(Z^{-1})$ satisfying Eq. (4) with the trailing coefficient of $\hat{\mathbf{R}}(Z^{-1})$ in upper echelon form.[3,6] Reversing the coefficients in this case yields a weak Popov form.

Starting from $\mathbf{M}(Z) = \mathbf{I}_m$ and $\vec{\omega} = \vec{0}$, the FFreduce algorithm computes order bases for increasing $\vec{\omega}$ until the desired order is reached. The recursion formulas to increase the order of an order basis from $\vec{\omega}$ to $\vec{\omega} + \vec{e}_j$ are given by the following theorem (see Theorem 6.1 in Ref. 3):

Theorem 2.1. *Let* $\mathbf{M}(Z)$ *be an order basis of order* $\vec{\omega}$ *and degree* $\vec{\mu}$, *and* $r_\ell = c_{\omega_j}((\mathbf{M}(\mathbf{Z}) \cdot \mathbf{F}(Z))_{\ell,j})$. *If* $r_\ell = 0$ *for all* $\ell = 1, \ldots, m$, *then* $\mathbf{M}(Z)$ *is an order basis of order* $\vec{\omega} + \vec{e}_j$ *and degree* $\vec{\mu}$. *Otherwise, we choose a pivot* π *such that*

$$\pi = \min_{1 \le \ell \le m} \left\{ \ell : r_\ell \neq 0, \mu_\ell = \min_{1 \le j \le m} \{\mu_j : r_j \neq 0\} \right\}, \tag{6}$$

and let $p_\ell = c_{\mu_\ell - 1 + \delta_{\pi,\ell}}(\mathbf{M}(Z)_{\pi,\ell})$. *Then an order basis* $\widetilde{\mathbf{M}}(Z)$ *of order* $\vec{\omega} + \vec{e}_j$ *and degree* $\vec{\mu} + \vec{e}_\pi$ *can be computed by*

$$p_\pi \cdot \widetilde{\mathbf{M}}(Z)_{\ell,*} = r_\pi \cdot \mathbf{M}(Z)_{\ell,*} - r_\ell \cdot \mathbf{M}(Z)_{\pi,*} \quad \text{for } \ell \neq \pi; \tag{7}$$

$$\sigma(p_\pi) \cdot \widetilde{\mathbf{M}}(Z)_{\pi,*} = (r_\pi \cdot Z - \delta(r_\pi)) \cdot \mathbf{M}(Z)_{\pi,*} - \sum_{\ell \neq \pi} \sigma(p_\ell) \cdot \mathbf{M}(z)_{\ell,*}. \tag{8}$$

When the coefficients of the Ore polynomials come from an integral domain such as $\mathbb{Z}[t]$, no fraction is introduced while applying Eq. (7) and Eq. (8). We also note that the degree of the order basis computed depends on the pivot choices in Eq. (6) and cannot be predicted in advance.

3. Linear Algebra Formulation

Given row degree $\vec{\mu}$ and order $\vec{\omega}$, the coefficients in the order basis $\mathbf{M}(Z)$ can be viewed as a solution to a linear system of equations over the coefficient ring. By equating the coefficients of like powers, each row of the order basis satisfies a system of equations of the form

$$\left[\cdots \left| \begin{array}{c} Z^0 \cdots Z^{\mu_k - 1 + \delta_{1,k}} \\ p_k^{(0)} \cdots p_k^{(\mu_k - 1 + \delta_{1,k})} \end{array} \right| \cdots \right] \cdot \begin{bmatrix} & Z^0 & \cdots & Z^{\vec{\omega}-\vec{e}} & \\ & \vdots & & & \\ \cdots & Z^0 \cdot F_{k,\cdot}(Z) & \cdots \\ & \vdots & & & \\ \cdots & Z^{\mu_k - 1 + \delta_{1,k}} \cdot F_{k,\cdot}(Z) & \cdots \\ & \vdots & & & \end{bmatrix} = \mathbf{0}. \tag{9}$$

More formally, for any $\mathbf{P}(Z) \in \mathbb{Q}[t][Z]^{m \times n}$ we define

$$\mathbf{P}_{\vec{v}} = \left[P_{*,1}^{(0)} \cdots P_{*,1}^{(v_1)} | \cdots | P_{*,n}^{(0)} \cdots P_{*,n}^{(v_n)} \right]. \tag{10}$$

We also define (recall that $\vec{\omega} = \sigma \cdot \vec{e}$)

$$
K(\vec{\mu}, \vec{\omega}) = \begin{bmatrix}
c_0(& \mathbf{F}(Z)_{1,*}) & \cdots & c_{\sigma-1}(& \mathbf{F}(Z)_{1,*}) \\
& \vdots & & & \vdots \\
c_0(Z^{\mu_1} \cdot & \mathbf{F}(Z)_{1,*}) & \cdots & c_{\sigma-1}(Z^{\mu_1} \cdot & \mathbf{F}(Z)_{1,*}) \\
& \vdots & & & \vdots \\
c_0(& \mathbf{F}(Z)_{m,*}) & \cdots & c_{\sigma-1}(& \mathbf{F}(Z)_{m,*}) \\
& \vdots & & & \vdots \\
c_0(Z^{\mu_m} \cdot & \mathbf{F}(Z)_{m,*}) & \cdots & c_{\sigma-1}(Z^{\mu_m} \cdot & \mathbf{F}(Z)_{m,*})
\end{bmatrix}. \tag{11}
$$

Then the i-th row of the order basis satisfies

$$
(\mathbf{M}_{i,*})_{\vec{\mu}-\vec{e}+\vec{e}_i} \cdot K(\vec{\mu} - \vec{e} + \vec{e}_i, \vec{\omega}) = \mathbf{0}. \tag{12}
$$

The matrix $K(\vec{\mu}, \vec{\omega})$ has dimensions $|\vec{\mu} + \vec{e}| \times |\vec{\omega}|$, and is called a *striped Krylov matrix* (with m stripes). This is a generalization of the well-known Sylvester matrix when $m = 2$ and $n = 1$.

Example 3.1. Let $\vec{\mu} = (2, 2)$, $\vec{\omega} = (3, 3)$, and

$$
\mathbf{F}(Z) = \begin{bmatrix} 2Z^2 + 3tZ + 6t^2 & Z^2 - Z + 2 \\ (t-1)Z + 3t^3 & 3tZ + t \end{bmatrix} \in \mathbb{Z}[t][Z; \sigma, \delta]^{2\times2}, \tag{13}
$$

with $\sigma(a(t)) = a(t)$ and $\delta(a(t)) = \frac{d}{dt}a(t)$. Then

$$
K(\vec{\mu}, \vec{\omega}) = \left[\begin{array}{cc|ccc|cc}
6t^2 & 2 & 3t & -1 & & 2 & 1 \\
12t & 0 & 6t^2 + 3 & 2 & & 3t & -1 \\
12 & 0 & 24t & 0 & & 6t^2 + 6 & 2 \\
\hline
3t^3 & t & t-1 & 3t & & 0 & 0 \\
9t^2 & 1 & 3t^3 + 1 & t+3 & & t-1 & 3t \\
18t & 0 & 18t^2 & 2 & & 3t^3 + 2 & t + 6
\end{array}\right]. \tag{14}
$$

We also define the matrix $K^*(\vec{\mu}, \vec{\omega}) = K(\vec{\mu}, \vec{\omega})_{*,J}$ where J is the lexicographically smallest set of column indices such that $K^*(\sigma n \cdot \vec{e}, \vec{\omega})_{*,J}$ has full column rank (this is called the *rank profile* in Ref. 16). An order basis of degree $\vec{\mu}$ and order $\vec{\omega}$ exists if $K^*(\vec{\mu} - \vec{e}, \vec{\omega})$ is nonsingular,[4] and in that case, Eq. (12) has a solution space of dimension one. The system can be transformed into the following system with a unique solution:

$$
(\mathbf{M}_{i,*})_{\vec{\mu}-\vec{e}} \cdot K^*(\vec{\mu} - \vec{e}, \vec{\omega}) = d \cdot \left[c_0\left(Z^{\mu_i} \cdot \mathbf{F}(Z)_{i,*}\right) \cdots c_{\sigma-1}\left(Z^{\mu_i} \cdot \mathbf{F}(Z)_{i,*}\right)\right] \tag{15}
$$

where $d = \pm \det K^*(\vec{\mu} - \vec{e}, \vec{\omega})$. In other words, we are interested in the Cramer solution of Eq. (15). Thus, the elements of the solution can be written as determinants of submatrices of $K(\vec{\mu}, \vec{\omega})$. If $\mathbf{F}(Z) \in \mathbb{Z}[t][Z]^{m \times n}$, then the solution has entries in $\mathbb{Z}[t][Z]$ as well.

The FFreduce algorithm performs fraction-free elimination of the matrix $K(\vec{\mu}, \vec{\omega})$ efficiently by taking advantage of the inherent structure in the matrix $K(\vec{\mu}, \vec{\omega})$. Indeed the algorithm has the effect of maintaining only one row in each of the m stripes. It can be shown that p_π is the pivot in the previous step in computing $\mathbf{M}(Z)$, in a way similar to fraction-free Gaussian elimination of Bareiss.[1] The elements of $\mathbf{M}(Z)$ are Cramer solutions to Eq. (15). The order basis $\mathbf{M}(Z)$ can be viewed as the transformation matrix representing the row operations performed during fraction-free Gaussian elimination.

Example 3.2. Let $\mathbf{F}(Z)$ be defined as in Example 3.1, so that we are performing Gaussian elimination on $K(\cdot, \cdot)$. In the first step of FFreduce, we choose $\pi = 1$ to eliminate the first column so that the row considered in the first stripe is advanced to the second row of the stripe. This gives

$$\mathbf{M}(Z) = \begin{bmatrix} 6t^2 Z - 12t & 0 \\ -3t^3 & 6t^2 \end{bmatrix}. \tag{16}$$

We omit giving the residual explicitly due to coefficient growth. In the second step, we again choose $\pi = 1$ to eliminate the second column. Note that here, the first row in the second stripe is zero in the second column, even though the remaining rows in the stripe have nonzero entries in the column. The choice of pivot depends only on the current row of each stripe. This gives an order basis of order $(1,1)$ and degree $(2,0)$:

$$\mathbf{M}(Z) = \begin{bmatrix} -24tZ^2 + 24Z & 0 \\ 12t^2 & -24t \end{bmatrix}. \tag{17}$$

For the third column, the entry of the current row in the second stripe is nonzero, so that we choose $\pi = 2$ as the pivot and obtain an order basis of order $(2,1)$ and degree $(2,1)$. Thus, the first three elements of J are 1, 2, and 3. Note that at each step, we are solving a linear system whose coefficient matrix grows column-wise (as the order increases) and row-wise (as the degree of the order bases increases).

As we can see, the row degrees of the order bases computed correspond to the number of times a row in each stripe is used as a pivot in the elimination process. The column indices in J give the columns in which elimination is performed (i.e. there is a nonzero entry in the current row of one of the stripes).

If we perform the same computation in $\mathbb{Z}_3$, we see that the first column is zero and we will reach an order basis of order $(2,1)$ and degree $(1,1)$. Here, the first two elements of J are 2 and 3. On the other hand, performing the computation in $\mathbb{Z}_2$ results in an order basis of order $(2,1)$ and degree $(1,1)$, with the first two elements of J being 1 and 3. This shows that when a prime is "unlucky", the rank profile (and hence the set of column indices J) as well as the sequence of row degrees of order bases constructed (the computational path) can be different from the correct ones.

4. Reduction to $\mathbb{Z}_p[t][Z]$

In this section, we show how techniques used in modular algorithms for polynomial matrices[7] can be extended to Ore polynomial matrices for the reduction of $\mathbb{Z}[t][Z]$ to $\mathbb{Z}_p[t][Z]$. We omit proofs which are similar to the polynomial matrix case and instead only highlight the differences. We refer the reader to Ref. 7 for more details.

Given the Ore polynomial ring $\mathbb{Z}[t][Z; \sigma, \delta]$ and a prime p, we define the modular homomorphism ϕ_p by

$$\phi_p(A) = A \bmod p, \quad \text{for } A \in \mathbb{Z}[t][Z; \sigma, \delta]. \tag{18}$$

This gives a mapping from the Ore polynomial ring $\mathbb{Z}[t][Z; \sigma, \delta]$ to the ring $\mathbb{Z}_p[t][Z; \sigma_p, \delta_p]$, where

$$\sigma_p(\phi_p(f)) = \phi_p(\sigma(f)) \text{ and } \delta_p(\phi_p(f)) = \phi_p(\delta(f)) \text{ for } f \in \mathbb{Z}[t] \ . \tag{19}$$

We are interested in reducing the computation over $\mathbb{Z}[t][Z; \sigma, \delta]$ to computations over $\mathbb{Z}_p[t][Z; \sigma_p, \delta_p]$ for a certain number of primes p. The results are then combined using Chinese remaindering to obtain the desired result in $\mathbb{Z}[t][Z; \sigma, \delta]$. Three issues need to be addressed: normalization of the results in $\mathbb{Z}_p[t][Z; \sigma_p, \delta_p]$, the detection of unlucky primes, and a termination criteria.

The main idea is that the computation of an order basis can be viewed as the computation of solutions to linear systems of the form in Eq. (15). As illustrated in Example 3.2, $r_\ell = 0$ for all $\ell = 1, \ldots, m$ in any one step implies that the corresponding column in the matrix $K(\cdot, \cdot)$ is linearly dependent. Otherwise, the choice of π refers to pivot row used in the elimination. Therefore, in order to be sure that the results computed in $\mathbb{Z}_p[t][Z; \sigma_p, \delta_p]$ is the same as those computed in $\mathbb{Z}[t][Z; \sigma, \delta]$, we need to ensure that the same systems of equations are solved (i.e. the rows and columns appearing in $K^*(\cdot, \cdot)$ is the same) and that the sign of d in Eq. (15) is chosen consistently across all primes p.

The sign of d can be computed consistently by applying the following formula together with the recursion formulas in Eq. (7) and Eq. (8):

$$\epsilon_0 = 1$$

$$\epsilon_{k+1} = \begin{cases} \epsilon_k & \text{if } r_\ell = 0 \text{ for all } \ell, \\ \epsilon_k \cdot (-1)^{\sum_{i=\pi_k+1}^{m} (\mu_k)_i} & \text{otherwise,} \end{cases} \tag{20}$$

where π_k and $\vec{\mu}_k$ are the values of π and $\vec{\mu}$ at the k-th step of the algorithm (Lemma 5.1(c) in Ref. 4). Multiplying the results obtained using Eq. (7) and Eq. (8) by ϵ_k at each step, this normalization ensures that $d = \det K^*(\vec{\mu} - \vec{e}, \vec{\omega})$. If the results in $\mathbb{Z}_p[t][Z; \sigma_p, \delta_p]$ are computed by some other means (see Sec. 5), we simply need to ensure that the sign is consistent with Eq. (20).

4.1. Lucky Homomorphisms

Let $hc(f)$ be the leading coefficient of $f \in Z[t][Z; \sigma, \delta]$ in t (the head coefficient). If $\phi_p(hc(\sigma(t))) \neq 0$, then σ_p is an Ore ring homomorphism and $\mathbb{Z}_p[t][Z; \sigma_p, \delta_p]$ is an Ore polynomial ring.[11,13] Thus, all primes p such that $\phi_p(hc(\sigma(t))) = 0$ are immediately discarded.

Let $\vec{\mu}$ and J be the degrees and index set computed for an order basis of order $\vec{\omega}$ in $Z[t][Z; \sigma, \delta]$, and $\vec{\mu}_p$ and J_p be those computed in $\mathbb{Z}_p[t][Z; \sigma_p, \delta_p]$. Also, let d be the constant in Eq. (15) computed in $Z[t][Z; \sigma, \delta]$. To ensure that the same systems of equations is solved in Eq. (15), we must ensure that $(\vec{\mu}, J) = (\vec{\mu}_p, J_p)$. This can be defined formally as follows.

Definition 4.1. The homomorphism ϕ_p is *lucky* if

(1) $\phi_p(d) \neq 0$;
(2) $|\vec{\mu}| = |\vec{\mu}_p|$; and
(3) $\phi_p(hc(\sigma(t))) \neq 0$.

Otherwise, it is *unlucky*.

It can be shown that if ϕ_p is lucky, then $(\vec{\mu}, J) = (\vec{\mu}_p, J_p)$.[7] In that case, the order basis and residual computed in $\mathbb{Z}_p[t][Z; \sigma_p, \delta_p]$ are images of the desired results under ϕ_p. Moreover, if p is unlucky then it must divide $hc(\sigma(t))$ or a minor of $K(\vec{\mu}, \vec{\omega})$, so the number of unlucky homomorphisms is finite.[7] More precisely, if κ is a bound on the coefficients of $Z^\ell \cdot \mathbf{F}(i, j)$ for $0 \leq \ell \leq mN + 1$, then the number of unlucky homomorphisms is at most

$$\max\left(\log_2 hc(\sigma(t)), (mN + 1)^2 n^2 \log_2\left(\kappa\sqrt{(mN + 1)n}\right)\right) \tag{21}$$

as we are only interested in minors of size up to $(mN + 1)n$. In practice, however, this is a very pessimistic estimate and unlucky homomorphisms are rarely encountered.

We note that the sequence of row degrees of the order bases constructed during the FFreduce algorithm represents the choice of pivots (see Example 3.2). We call this sequence a *computational path*. If we define the path $W = \{\vec{w}_k\}_{k=0,1,2,\ldots}$ by $\vec{w}_0 = \vec{0}$ and $\vec{w}_{k+1} = \vec{w}_k + \vec{e}_{k \bmod m+1}$, then W is the sequence of degrees followed by the FFreduce algorithm if $r_\ell \neq 0$ for all ℓ at every step. It was shown that the final degree $\vec{\mu}$ is the unique closest *normal point* to w (see Theorem 7.3 in Ref. 4). That is, if $K^*(\vec{v} - \vec{e}, \vec{w})$ is nonsingular for some $\vec{v}$ such that $|\vec{v}| = |\vec{\mu}|$, then

$$|\max(\vec{0}, \vec{w}_k - \vec{\mu})| \leq |\max(\vec{0}, \vec{w}_k - \vec{v})| \text{ for all } k \geq 0. \tag{22}$$

Since $\vec{\mu}$ and J are not known a priori, we need criteria to compare the results computed under two homomorphisms and determine if one of them is unlucky. This is similar to the case for polynomial matrices.[7] This allows one to incrementally compute homomorphic images and detect unlucky homomorphisms by comparing the current image against the previous ones.

Theorem 4.1. *Suppose $\phi_p(hc(\sigma(t))) \neq 0$, $\phi_p(F^{(0)}) \neq \mathbf{0}$, and $\phi_q(F^{(0)}) \neq \mathbf{0}$. Then ϕ_p is unlucky if one of the following holds:*

(1) $|\vec{\mu}_p| = |\vec{\mu}_q|$ and $J_p >_{\text{lex}} J_q$;
(2) $|\vec{\mu}_p| = |\vec{\mu}_q|$, $J_p = J_q$, and $\vec{\mu}_q$ is closer to W than $\vec{\mu}_p$;
(3) $|\vec{\mu}_p| < |\vec{\mu}_q|$.

We remark that this theorem is used only to detect unlucky homomorphisms and cannot be used to detect lucky ones.

Assuming that we can compute the order basis $\mathbf{M}_p(Z)$ (and the corresponding row degree $\vec{\mu}_p$ and column index set J_p) for $\phi_p(F(Z))$ over $\mathbb{Z}_p[t][Z; \sigma_p, \delta_p]$, Theorem 4.1 allows the results computed under two different homomorphisms be compared to detect unlucky homomorphisms.

4.2. *Termination*

As all coefficients in $\mathbf{M}(Z)$ are Cramer solutions of the linear systems in Eq. (15), they can be written as determinants of the coefficient matrices.[4] Using Hadamard's inequality, we can bound the size of the coefficients and terminate the modular algorithm when the product of the moduli exceeds this bound. Unfortunately, the Hadamard bound can be extremely pessimistic. A common approach is to reconstruct the results incrementally one prime at a time using Garner's algorithm,[8] and verify the results when they do not change for a certain number (e.g. one) of additional homomorphisms. The verification step typically involves trial division or multiplication.

An approach of Cabay[5] was first used for solving systems of linear equations with a modular algorithm without the need for verification. This was extended to the case of polynomial matrix normal form computations.[7] This technique can be extended to Ore polynomial matrices.

Theorem 4.2. Let $\|f(t)\|_\infty = \max_i |f^{(i)}|$ where $f(t) = \sum_i f^{(i)} t^i$. Suppose that for all $1 \le i \le m$, $1 \le j \le n$, $0 \le k < mN + 1$, and $0 \le \ell \le mN + 1$, we have

$$deg_t \left(c_k \left(Z^\ell \cdot \mathbf{F}(Z)_{i,j} \right) \right) \le T; \tag{23}$$

$$\left\| c_k \left(Z^\ell \cdot \mathbf{F}(Z)_{i,j} \right) \right\|_\infty \le \kappa. \tag{24}$$

Suppose the primes are ordered such that $p_1 < p_2 < \cdots$, and that

$$((mN + 1)n)\kappa(T + 1) \le p_1 \cdots p_\tau \tag{25}$$

for all $j = 1, \ldots, n$. If $\widetilde{\mathbf{M}}(Z)$ and $\widetilde{\mathbf{R}}(Z)$ are the reconstructed results in the modular algorithm and have not changed for τ additional primes, then $\widetilde{\mathbf{M}}(Z)$ and $\widetilde{\mathbf{R}}(Z)$ give a solution to Eq. (4).

Proof. The proof of this theorem follows that of Theorem 6.1 in Ref. 7. We note that the coefficients of $\mathbf{M}(Z)$ are solutions to the linear systems of equations in Eq. (15), where the size of the matrix $K^*(\vec{\mu} - \vec{e}, \vec{\omega})$ is $(mN + 1)n$. The only difference here is that we need to note that by Hadamard's inequality, $\deg_t(f) \le (mN + 1)nT$ where f is a coefficient in $\mathbf{M}(Z)$, and that $\|fg\|_\infty \le (\min(\deg_t(f), \deg_t(g)) + 1)\|f\|_\infty \|g\|_\infty$. $\square$

The early termination criteria is most useful if τ is small. In particular, if $\tau = 1$ then the proposed criteria is clearly an improvement over a traditional modular algorithm. This is often true in practical cases.[7] In the worst case, the termination condition of Eq. (25) is satisfied when the traditional Hadamard bound is reached. Thus, the early termination strategy is no worse than the traditional one.

Remark 4.1. The bound in Eq. (25) is based on the norm of a column in $K^*(\vec{\mu} - \vec{e}, \vec{\omega})$. In specific cases where σ and δ are known, the bound can be refined.

5. Reduction to $\mathbb{Z}_p$

While we do not suffer coefficient growth in the size of the integer coefficients when solving the problems in $\mathbb{Z}_p[t][Z; \sigma_p, \delta_p]$, the growth of the degrees of the coefficients with respect to t still needs to be controlled. In this section, we show how the computation of an order basis in $\mathbb{Z}_p[t][Z; \sigma_p, \delta_p]$ can be transformed into a number of linear algebra problems over $\mathbb{Z}_p$.

It has been shown that the evaluation homomorphisms $t \leftarrow \alpha$ are usually not Ore ring homomorphisms because Ore polynomial rings over $\mathbb{Z}_p$ must be commutative.[11,13] As a result, it is not possible to apply the same technique in the previous section to obtain a modular algorithm to compute order bases in $\mathbb{Z}_p[t][Z; \sigma_p, \delta_p]$ by reducing the problem into ones with coefficients in $\mathbb{Z}_p[Z]$. However, when the problem in $\mathbb{Z}_p[t][Z; \sigma_p, \delta_p]$ is viewed as the linear algebra problem in Eq. (15) over $\mathbb{Z}_p[t]$, each entry in the striped Krylov matrix can be reduced by an evaluation homomorphism to obtain a number of linear algebra problems in $\mathbb{Z}_p$. This is essentially the approach taken to compute a GCRD and a LCLM of Ore polynomials[11,13]—the image of the coefficient matrix over $\mathbb{Z}_p$ is constructed and Gaussian elimination is performed on this matrix over $\mathbb{Z}_p$.

In the Ore polynomial case, the coefficient matrix in Eq. (15) is the well-known Sylvester matrix and the computational path corresponds directly to the degree sequence of the polynomial remainder sequence.[8] In the Ore polynomial matrix case, however, we do not know of such a correspondence and neither the computational path nor the final configuration of the striped Krylov matrix is known a priori. Thus, it is not sufficient to simply construct the coefficient matrix and then operate on it.

Our modular algorithm to compute order bases in $\mathbb{Z}_p[t][Z; \sigma_p, \delta_p]$ consists of the following steps:

(1) Choose a number of evaluation points α.
(2) For each evaluation point, apply the evaluation homomorphism $t \leftarrow \alpha$ and solve the reduced linear algebra problem over $\mathbb{Z}_p$.
(3) Combine the results obtained from lucky evaluation points by polynomial interpolation.

This is the traditional framework for modular algorithm. However, the application of the evaluation homomorphism and solution to the linear algebra problem is non-trivial.

5.1. Applying Evaluation Homomorphisms and Computation in $\mathbb{Z}_p$

Our approach can be seen as a generalization of the modular algorithm for Ore polynomials.[11,13] Instead of constructing the entire coefficient matrix $K(\cdot, \vec{\omega})$ in $\mathbb{Z}_p$ and performing Gaussian elimination on this matrix, we will incrementally construct the striped Krylov matrix. Whenever we choose the pivot π in an elimination step, the row corresponding to $Z^{\mu_\pi+1} \cdot \mathbf{F}(Z)_{\pi,*}$ needs to be added to the striped Krylov matrix. Once the row is added, elimination must be performed so that the new row also has the required order. However, this is not completely straightforward as the next example shows.

Example 5.1. Let $\mathbf{F}(Z)$ be defined as in Example 3.1. Applying the evaluation $t \leftarrow 0$, the image of the striped Krylov matrix $K(\vec{\mu}, \vec{\omega})$ over $\mathbb{Z}_5$ is

$$
K = \begin{bmatrix}
0\ 2 & 0\ 4 & 2\ 1 \\
0\ 0 & 3\ 2 & 0\ 4 \\
2\ 0 & 0\ 0 & 1\ 2 \\
\hline
0\ 0 & 4\ 0 & 0\ 0 \\
0\ 1 & 1\ 3 & 4\ 0 \\
0\ 0 & 0\ 2 & 2\ 1
\end{bmatrix} . \tag{26}
$$

When the third row in the first stripe is finally added to the matrix, we notice that the first column is linearly independent. In fact, there is no need to perform elimination on the added row.

The sequences of pivot rows and columns over $\mathbb{Z}_5$ is different from those over $\mathbb{Z}_5[t]$, but the set of pivot rows and columns used at the end are the same in this case as K has full rank. After appropriate normalization, the solutions computed over $\mathbb{Z}_5$ are images of the corresponding solutions in $\mathbb{Z}_5[t]$.

We now describe the process by which the elimination is performed. The computation of quantities such as $\vec{\mu}$ and π are similar to the FFreduce algorithm; they will not be given explicitly here. We will keep the striped Krylov matrix K in the form of matrix of coefficients in $\mathbb{Z}_p$. The entries in the transformation matrix $\mathbf{T}(Z)$ will be kept in polynomial form to simplify the manipulations required when a new row is added. In addition, we maintain a matrix $\mathbf{C}(Z)$ where $\mathbf{C}(Z)_{i,*} = Z^{\mu_i} \cdot \mathbf{F}(Z)_{i,*} \in \mathbb{Z}_p[t][Z; \sigma_p, \delta_p]^{1 \times n}$. The vector $\mathbf{C}(Z)_{i,*}$ represents the last row of stripe i added to K and allows us to quickly add the next row.

The algorithm proceeds as follows.

(1) Initially, set $\mathbf{C}(Z) = \mathbf{F}(Z)$. Also initialize K to include the m rows consisting of coefficients of $\mathbf{F}(Z)_{i,*}$ up to order $\vec{\omega}$ evaluated at $t = \alpha$, and a transformation matrix $\mathbf{T}(Z) = I_m$.

(2) For each elimination step, the residuals r_i in Theorem 2.1 can be obtained by examining the appropriate row and column in K based on $\vec{\mu}$ and σ and the form of the striped Krylov matrix in Eq. (11). If $r_i \neq 0$ for some $1 \leq i \leq m$, add the step number j to set of column indices J. Once the pivot row is chosen, we apply standard row operations in $\mathbb{Z}_p$ (non-fraction-free) to eliminate all other rows of K for that column. We also need to add the next row for stripe π:

   (a) $\mathbf{C}(Z)_{\pi,*} \leftarrow Z \cdot \mathbf{C}(Z)_{\pi,*}$. This operation is performed in $\mathbb{Z}_p[t][Z; \sigma_p, \delta_p]$.

   (b) Evaluate $\mathbf{C}(Z)_{\pi,*}$ at $t = \alpha$. Add the coefficients as a new row to $\mathbf{K}$, and perform row interchanges so that the stripes in $\mathbf{K}$ are in the form in Eq. (11). Add a corresponding row to $\mathbf{T}(Z)$, whose only nonzero entry is $Z^{\mu_\pi + 1}$ in column π.

   (c) Perform row operations to eliminate the added row up to column j using all rows that have previously been used as pivots. The same row operations are applied to $\mathbf{T}(Z)$. During the reduction, if the added row introduces a new linearly independent column, add the column index to J and repeat steps 2a to 2c.

(3) When the row reductions are complete, let K^* be the triangular submatrix of $K_{*,J}$ consisting of all but the last added row for each stripe. Compute the determinant $d = \det K^*$ as a product of the diagonal elements. Adjust the sign of d based on the row interchanges performed (see Eq. (20)).

(4) The i-th row of the order basis $\mathbf{M}(Z)$ and the residual $\mathbf{R}(Z)$ can be extracted from the rows corresponding to $Z^{\mu_i} \cdot \mathbf{F}(Z)_{i,*}$ in $d \cdot \mathbf{T}(Z)$ and $d \cdot K$.

Although we have lost the ability of FFreduce to take advantage of the structure of the striped Krylov matrix, we gained the ability to control coefficient growth. Note that coefficient growth is not completely eliminated, since the computation of $\mathbf{C}(Z)$ may introduce growth in the degree in t when $Z^k \cdot \mathbf{F}(Z)$. However, the growth arising from Gaussian elimination is eliminated. Furthermore, the degree in t does not grow when multiplying by Z in many practical cases (see Remark 6.3).

Example 5.2. Continuing from Example 5.1, we first start with

$$K = \begin{bmatrix} 0\,2 & 0\,4 & 2\,1 \\ 0\,0 & 4\,0 & 0\,0 \end{bmatrix}. \tag{27}$$

Eliminating the second column with pivot $\pi = 1$ requires no row operation. The next row for the first stripe is added to obtain

$$K = \begin{bmatrix} 0\,2 & 0\,4 & 2\,1 \\ 0\,0 & 3\,2 & 0\,4 \\ 0\,0 & 4\,0 & 0\,0 \end{bmatrix}. \tag{28}$$

In the next step, we choose $\pi = 2$ to eliminate the third column. Adding the next row and eliminating gives

$$K = \begin{bmatrix} 0\,2 & 0\,4 & 2\,1 \\ 0\,0 & 0\,2 & 0\,4 \\ 0\,0 & 4\,0 & 0\,0 \\ 0\,0 & 0\,1 & 3\,2 \end{bmatrix}. \tag{29}$$

Next, we choose $\pi = 1$ to eliminate the fourth column. After adding the next row for the first stripe, we get:

$$K = \begin{bmatrix} 0\,2 & 0\,4 & 2\,1 \\ 0\,0 & 0\,2 & 0\,4 \\ 2\,0 & 0\,0 & 1\,2 \\ 0\,0 & 4\,0 & 0\,0 \\ 0\,0 & 0\,0 & 3\,0 \end{bmatrix}. \tag{30}$$

The new row introduces a new linearly independent column. The next row for the first stripe will be added immediately. Continuing in this manner gives the coefficients of the residual $\mathbf{R}(Z)$ as the last row of each stripe.

5.2. *Lucky Homomorphisms and Termination*

Since our definition of lucky homomorphisms is originally based on the linear algebra formulation of the order basis problem, the same definition can be easily applied to the reduction of the linear algebra problem from $\mathbb{Z}_p[t]$ to $\mathbb{Z}_p$. In particular, Definition 4.1 and Theorem 4.1 can be applied simply by changing ϕ_p and ϕ_q to the appropriate evaluation homomorphisms.

Similarly, the termination criteria is also originally based on the linear algebra formulation in Eq. (15). Again, we may apply Hadamard's inequality to obtain bounds on the size (degrees in t) of the coefficients in the solutions, but we prefer to have an early termination condition that is sensitive to the size of the output. Since we are now dealing with linear systems of equations, we can apply the technique of Cabay[5] directly after modifying the theorem to use the degree measure as the coefficient norm.

Theorem 5.1. *Suppose that for all* $1 \leq i \leq m$, $1 \leq j \leq n$, $0 \leq k < mN+1$, *and* $0 \leq \ell \leq mN + 1$, *we have*

$$deg_t \left(c_k \left(Z^\ell \cdot \mathbf{F}(Z)_{i,j} \right) \right) \leq T. \tag{31}$$

Suppose that $\widetilde{\mathbf{M}}(Z)$ *and* $\widetilde{\mathbf{R}}(Z)$ *are the reconstructed results in the modular algorithm and have not changed for* T *additional evaluation points. Then* $\widetilde{\mathbf{M}}(Z)$ *and* $\widetilde{\mathbf{R}}(Z)$ *give a solution to Eq. (4) in* $\mathbb{Z}_p[t][Z; \sigma, \delta]$.

6. Complexity Analysis

To compare the new modular algorithm against the fraction-free FFreduce algorithm, we need to give the complexity of FFreduce in our context, as the analysis in our previous work[3,6] are for general coefficient domains and can be refined. The proof is similar to the ones in our previous work and is omitted.

Theorem 6.1. *Suppose that for all* $1 \leq i \leq m$, $1 \leq j \leq n$, $0 \leq k < mN+1$, *and* $0 \leq \ell \leq mN + 1$, *we have*

$$deg_t \left(c_k \left(Z^\ell \cdot \mathbf{F}(Z)_{i,j} \right) \right) \leq T; \tag{32}$$

$$\left\| c_k \left(Z^\ell \cdot \mathbf{F}(Z)_{i,j} \right) \right\|_\infty \leq \kappa. \tag{33}$$

Then an order basis and a residual of order $\vec{\omega} = (mN+1) \cdot \vec{e}$ *over* $\mathbb{Z}[t][Z; \sigma, \delta]$ *can be computed in* $O((m+n)(mnN)^4 T^2 \mathbf{M}(mnN(\log(T\kappa))))$ *bit operations by the FFreduce algorithm, where* $O(\mathbf{M}(k))$ *is the complexity of multiply two* k-*bit integers.*

We now analyze the complexity of our algorithm. We first examine the complexity of the computation of the order basis over $\mathbb{Z}_p[t][Z; \sigma_p, \delta_p]$.

Theorem 6.2. *Let* T *be a bound on the degree of the coefficients in* t, *such that*

$$deg_t \left(c_k \left(Z^\ell \cdot \mathbf{F}(Z)_{i,j} \right) \right) \leq T \tag{34}$$

for $1 \leq i \leq m$, $1 \leq j \leq n$, $0 \leq k < mN + 1$, *and* $0 \leq \ell \leq mN + 1$. *We also assume that two polynomials in* $\mathbb{Z}_p[t]$ *of degree* d *can be multiplied in* $O(d \log d)$ *operations in* $\mathbb{Z}_p$. *Then an order basis and a residual of order* $\vec{\omega} = (mN + 1) \cdot \vec{e}$ *over* $\mathbb{Z}_p[t][Z; \sigma_p, \delta_p]$ *can be computed in* $O((mnN)^3 mT(nN + T(\log mnNT)^2))$ *operations in* $\mathbb{Z}_p$.

Proof. Since the dimensions of the coefficient matrix in the system of equations in Eq. (15) is $(mN+1)n \times (mN+1)n$, it follows that the entries of the order basis and the residual have degrees in t bounded by $(mN+1)nT$ as they can be written as determinants of submatrices of the coefficient matrix. This implies that $O(mnNT)$ lucky evaluation points are needed. For each evaluation point, the evaluation can be done in $O((mnN)^2T)$ operations in $\mathbb{Z}_p$. The elimination can be done in $O((mnN)^3)$ operations. Finally, each coefficient can be interpolated in $O(\mathbf{PM}(mnNT)\log(mnNT))$ operations,[17] and we have $O(m^2nN)$ nonzero coefficients in $\mathbf{M}(Z)$ and $O(mnN)$ nonzero coefficients in $\mathbf{R}(Z)$, where $\mathbf{PM}(d)$ is the complexity for multiplying two degree d polynomials. The desired result now follows from the assumption on polynomial multiplication. □

Remark 6.1. In the analysis we have ignored the occurrences of unlucky homomorphisms as they rarely occur in practice. Also, we assumed that p is chosen large enough such that polynomial multiplication for degree d polynomials can be performed in $O(d \log d)$ operations.

We are now ready to give the complexity of the complete algorithm.

Theorem 6.3. *Suppose that for all $1 \leq i \leq m$, $1 \leq j \leq n$, $0 \leq k < mN+1$, and $0 \leq \ell \leq mN + 1$, we have*

$$deg_t \left(c_k \left(Z^\ell \cdot \mathbf{F}(Z)_{i,j}\right)\right) \leq T; \tag{35}$$

$$\left\|c_k \left(Z^\ell \cdot \mathbf{F}(Z)_{i,j}\right)\right\|_\infty \leq \kappa. \tag{36}$$

Then an order basis and a residual of order $\vec{\omega} = (mN+1)\cdot\vec{e}$ over $\mathbb{Z}[t][Z; \sigma, \delta]$ can be computed in

$$O(m(mnN)^3T(nN + T(\log mnNT)^2 + \log\log T\kappa)\mathbf{M}(mnN \log T\kappa))$$

bit operations by our modular algorithm.

Proof. By Hadamard's inequality, Eq. (15), and the inequality $\|fg\|_\infty \leq (\min(\deg_t(f), \deg_t(g)) + 1)T\|f\|_\infty\|g\|_\infty$, we see that $\left\|c_k \left(\mathbf{M}(\mathbf{Z})_{i,j}\right)\right\|_\infty$ and $\left\|c_k \left(\mathbf{R}(\mathbf{Z})_{i,j}\right)\right\|_\infty$ have $O(mnN \log T\kappa)$ bit length. Choosing primes of size $O(\log T\kappa)$, it follows that $O(mnN)$ lucky primes are needed. For each prime, the reduction modulo p can be done in $O(mnNT)$ operations in $\mathbb{Z}_p$, and the computation of the order basis and residual in $\mathbb{Z}_p[t][Z; \sigma, \delta]$ can be done in $O((mnN)^3mT(nN+T(\log mnNT)^2))$ operations in $\mathbb{Z}_p$ by Theorem 6.2. We note that each operation can be performed in $O(\mathbf{M}(\log T\kappa))$ bit operations.

Finally, there are $O(m^2 nN)$ nonzero coefficients in $\mathbf{M}(Z)$ and $O(mnN)$ nonzero coefficients in $\mathbf{R}(Z)$, each of which has degree $O(mnNT)$ in t. Each coefficient can be reconstructed in $O(\mathbf{M}(mnN \log T\kappa) \log(mnN \log T\kappa))$ bit operations by Chinese remaindering.[17]   □

Remark 6.2. We see that the complexity of the modular algorithm improves on the complexity of the fraction-free algorithm, but the advantage of the reduced coefficient growth is offset by the use of a larger striped Krylov matrix. In the analysis we have used the worst case bound that the number of rows in the striped Krylov matrix can be a factor of nN greater than that of $\mathbf{R}(Z)$ and $\mathbf{M}(Z)$ in Ore polynomial matrix form. In practice, however, the striped Krylov matrix usually never grows to its full size because there can be many zero rows, or because we can terminate the elimination earlier in the case of matrices of shift polynomials. The advantage of the modular algorithm is more significant in these cases.

Remark 6.3. In many practical applications, $\deg_t(\sigma(a)), \deg_t(\delta(a)) \leq \deg_t(a)$ for all $a \in \mathbb{Z}[t]$. In these cases, one can simplify Eq. (35) to

$$\deg_t \left(c_k \left(\mathbf{F}(Z)_{i,j} \right) \right) \leq T. \tag{37}$$

7. Implementation Considerations and Experimental Results

The modular algorithm has been implemented in Maple 9.5. Although the modular algorithm has a better complexity than the fraction-free algorithm, the modular algorithm has a larger overhead especially for small inputs. As a result, a careful implementation is needed for the modular algorithm to perform better than the fraction-free algorithm on inputs of reasonable size. We list below some optimizations used. Although they are straightforward, these optimizations have significant effect on the running time of the algorithm.

- Memory allocation and deallocation can be a significant overhead in the algorithm. To minimize memory management, we allocate one large matrix K for each prime p and use the same matrix for different evaluation points. There is also no need to reallocate the matrix when a new row is added, as long as the original matrix is large enough. In addition, we may augment the matrix K with the identity matrix to represent the coefficients of the order basis.

- When a new row is added to K, we do not perform row interchanges to maintain the coefficient matrix K in striped Krylov matrix form. Instead, we simply add the new row to the bottom and maintain a list of row indices to refer to the correct row. This reduces data movements in the algorithm.
- The computations on $\mathbf{C}(Z)$ for keeping track of the next row to be added to K are identical for all evaluation points α under $\mathbb{Z}_p$. Therefore, we compute the sequence of $\mathbf{C}(Z)$ only once for each prime p and reuse the results for all evaluation points.
- The `LinearAlgebra:-Modular` package is used to efficiently perform linear algebra operations in $\mathbb{Z}_p$.

We compare the performance of the modular algorithm and the fraction-free algorithm (FFreduce) below. The experiments were performed on a computer with a Xeon 2.70GHz processor and 16GB of RAM. In the experiments, we generated random matrices of Ore polynomials such that $\sigma(a(t)) = a(t)$ and $\delta(a(t)) = \frac{d}{dt}a(t)$. The results are given in Table 1 and Table 2. As expected, we see that the modular algorithm performs better than the fraction-free algorithm as the input size increases. Also, we show that using rational arithmetic in $\mathbb{Q}[t]$ to perform row reduction is impractical.

Table 1.   Comparison between modular and fraction-free algorithms for various input sizes ($\kappa = 5$, $T = 1$). Also shown are timings when the rational arithmetic in $\mathbb{Q}[t]$ is used (n/a means no result is computed after 7200 seconds).

m, n	N	FFreduce (s)	Modular (s)	Ratio	Rational (s)
2	1	0.037	0.178	0.208	3.073
2	2	0.089	0.278	0.320	n/a
2	4	1.453	3.226	0.449	n/a
2	8	21.984	26.789	0.821	n/a
2	16	75.048	118.796	0.630	n/a
3	1	0.564	1.150	0.490	n/a
3	2	3.928	6.032	0.652	n/a
3	4	64.498	55.991	1.150	n/a
3	8	401.708	339.383	1.190	n/a
4	2	54.213	46.776	1.160	n/a
4	4	1018.963	589.687	1.730	n/a
4	6	5107.881	3123.530	1.640	n/a
5	2	564.498	289.937	1.940	n/a
5	4	7770.646	4554.689	1.710	n/a
8	1	2434.662	1182.519	2.060	n/a
10	1	17124.143	11146.557	1.540	n/a

Table 2. Comparison between modular and fraction-free algorithms for various input sizes ($\kappa = 5$, $T = 2$).

m, n	N	FFreduce (s)	Modular (s)	Ratio
2	2	0.496	1.848	0.268
2	4	5.618	11.368	0.493
2	8	95.759	111.925	0.855
2	16	1723.170	1709.175	1.010
3	2	16.488	24.198	0.682
3	4	330.956	291.985	1.130
3	6	1373.799	1225.103	1.110

Although the modular algorithms is faster than the fraction-free algorithm for larger inputs, there were instances where the ratio between the running times of the fraction-free algorithm and the modular algorithm decreases slightly when the input size is increased. In these cases, the larger inputs actually gave output that are smaller than expected because the rank of the striped Krylov matrix is not full. Hence, the recurrence formulas in Eq. (7) and Eq. (8) are not applied as often, resulting in less coefficient growth in the output. Also, for larger inputs the overhead of garbage collection for the results computed under different evaluation points and different primes become more important. This is confirmed by observing that the running time for the modular algorithm improves significantly when the frequency of garbage collection is reduced[b]. A decision has to be made based on available memory in order to reduce running time while using a reasonable amount of memory.

8. Concluding Remarks

In this paper, we showed how to design an output-sensitive modular algorithm for performing row reductions on matrices of Ore polynomials. By examining the problem as a linear algebra problem in $\mathbb{Z}_p[t]$ and subsequently in $\mathbb{Z}_p$, we overcome the various issues in designing a modular algorithm—detection of unlucky homomorphisms, normalization, and termination. We have also shown that the modular algorithm is faster than the fraction-free algorithm for larger inputs both theoretically and experimentally.

A limitation in both the fraction-free and modular algorithms is that row reduction is performed on the low order terms. When the elements of the input matrix are shift polynomials, one may perform substitution to reverse

---

[b]Changing the kernel variable `gcfreq` to 5000000 from the default value of 1000000 reduces the running times by as much as 3 times in some cases.

the coefficients. However, this cannot be done for general Ore polynomials. We believe that it is possible to formulate row reduction based on high order terms as a nullspace computation, so that the modular algorithm described here can be applied to control coefficient growth for the computation of normal forms that are defined by leading coefficients such as the row-reduced form.

References

1. E. Bareiss. Sylvester's identity and multistep integer-preserving Gaussian elimination. *Math. Comp.*, 22:565–578, 1968.
2. B. Beckermann, H. Cheng, and G. Labahn. Fraction-free row reduction of matrices of skew polynomials. In *Proceedings of the 2002 International Symposium on Symbolic and Algebraic Computation*, pages 8–15. ACM, 2002.
3. B. Beckermann, H. Cheng, and G. Labahn. Fraction-free row reduction of matrices of Ore polynomials. *Journal of Symbolic Computation*, 41(5):513–543, 2006.
4. B. Beckermann and G. Labahn. Fraction-free computation of matrix rational interpolants and matrix GCDs. *SIAM J. Matrix Anal. and Appl.*, 22(1):114–144, 2000.
5. S. Cabay. Exact solution of linear equations. In *Proceedings of the Second Symposium on Symbolic and Algebraic Manipulation*, pages 392–398, 1971.
6. H. Cheng. *Algorithms for Normal Forms for Matrices of Polynomials and Ore Polynomials*. PhD thesis, University of Waterloo, 2003.
7. H. Cheng and G. Labahn. Output-sensitive modular algorithms for polynomial matrix normal forms. *To appear in Journal of Symbolic Computation*.
8. K. O. Geddes, S. R. Czapor, and G. Labahn. *Algorithms for Computer Algebra*. Kluwer Academic Publishers, 1992.
9. M. Giesbrecht, G. Labahn, and Y. Zhang. Computing valuation popov forms. In *Computer Algebra Systems and their Applications (CASA'05)*, 2005.
10. T. Kailath. *Linear Systems*. Prentice-Hall, 1980.
11. Z. Li. *A Subresultant Theory for Linear Differential, Linear Difference and Ore Polynomials, with Applications*. PhD thesis, RISC-Linz, Johannes Kepler University, Linz, Austria, 1996.
12. Z. Li. A subresultant theory for ore polynomials with applications. In *Proceedings of the 1998 International Symposium on Symbolic and Algebraic Computation*, pages 132–139. ACM, 1998.
13. Z. Li and I. Nemes. A modular algorithm for computing greatest common right divisors of ore polynomials. In *Proc. of the 1997 International Symposium on Symbolic and Algebraic Computation*, pages 282–289. ACM, 1997.
14. T. Mulders and A. Storjohann. On lattice reduction for polynomial matrices. *Journal of Symbolic Computation*, 35(4):377–401, 2003.
15. O. Ore. Theory of non-commutative polynomials. *Annals of Mathematics*, 34:480–508, 1933.
16. A. Storjohann. *Algorithms for Matrix Canonical Forms*. PhD thesis, Dept. of Computer Science, Swiss Federal Institute of Technology—ETH, 2000.
17. J. von zur Gathen and J. Gerhard. *Modern Computer Algebra*. Cambridge University Press, second edition, 2002.

BETA-EXPANSIONS OF PISOT AND SALEM NUMBERS

KEVIN G. HARE *

Department of Pure Mathematics
University of Waterloo
Waterloo, Ontario
Canada
N2L 3G1
kghare@math.uwaterloo.ca

This paper is based on a talk given at WWCA (Waterloo Workshop on Computer Algebra) held at Wilfird Laurier University, April 2006. This paper gives a history of beta-expansions, and surveys some of the computational aspects of beta-expansions. Special attention is given to how these beta-expansions relate to Pisot and Salem numbers. This paper also gives an overview of the computational issues that arose in the recent investigation of Allouche, Frougny and Hare,[2] upon which the talk at WWCA was based.

Keywords: Beta-expansions, Pisot numbers, Salem numbers, Univoque

1. Introduction and History

People have represented numbers in a varieties of different bases. By far the most common today is base 10, the current system inherited from the Arabics. Some other bases that have been used historically include the base 60 (mixed representation) of the Babylonians, and the base 20 (mixed representation) of the Mayans (Katz[31]). More recently, base 2 and 16 have become more common, with modern day computer science.

Let us consider the standard base 10 representation for now. Assume for convenience that $x \in [0, 1)$. We say

$$x = 0.a_1 a_2 a_3 \cdots = \sum_{i=1}^{\infty} \frac{a_i}{10^i}. \tag{1}$$

Here the $a_i \in \{0, 1, \cdots, 9\}$. The middle expression of Eq. (1) is called a base 10 representation of x. We say a representation of an $x \in [0, 1)$ is *eventually*

---

*Research of K. G. Hare supported, in part by NSERC of Canada *periodic* if it can be written as

$$x = 0.a_1a_2 \cdots a_k(a_{k+1}a_{k+2} \cdots a_n)^\omega.$$

If we assume further than $a_k \neq a_n$ and that the period has minimal length, then we say that the representation has a *pre-period* of length k and a *period* of length $(n - k)$. If the period has length 1 and $a_n = 0$, then we say that the representation is *finite* and we omit the 0^ω in the representation to get $x = 0.a_1a_2 \cdots a_k$.

An elementary result about decimal numbers is that the representation of a number is eventually periodic or finite if and only if it represents a rational number. It is easy to see that this is true for any base $q \in \mathbb{Z}$, with $q \geq 2$. An immediate question presents itself. What happens if $q \notin \mathbb{Z}$?

Rényi,[41] was among the first to study representations of a value x with a non-integer base β. These have come to be known as beta-expansions. To this end we define the beta-expansion of a number x.

Definition 1.1. Let $x \in \left[0, \frac{\lceil\beta\rceil-1}{\beta-1}\right)$ satisfy

$$x = \sum_{n=1}^{\infty} a_n\beta^{-n}$$

where $a_n \in \{0, 1, \cdots, \lceil\beta\rceil - 1\}$. Then $a_1a_2a_3 \cdots$ is a **beta-expansion** for x.

For the purposes of this paper, β is always a real non-integer, and q is always an integer. Furthermore, we always assume that β and q are strictly greater than 1. Note, for convenience, we omit the "0." in front of $a_1a_2 \cdots$. This is done first, to be consistent with the literature on beta-expansions, and secondly, because it is more natural to think of a beta-expansion as an infinite word, for which a number of the results have a more elegant description.

We now consider our first example with a non-integer base.

Example 1.1. Consider β the golden ratio, $\beta \approx 1.618$, the larger root of $\beta^2 - \beta - 1 = 0$.

Then

$$
\begin{aligned}
1 &= \tfrac{1}{\beta} + \tfrac{1}{\beta^2} && \text{or} \\
&= \tfrac{1}{\beta} + \tfrac{1}{\beta^3} + \tfrac{1}{\beta^4} && \text{or} \\
&= \tfrac{1}{\beta} + \tfrac{1}{\beta^3} + \tfrac{1}{\beta^5} + \tfrac{1}{\beta^7} + \tfrac{1}{\beta^9} + \cdots.
\end{aligned}
$$

So the beta-expansions of 1 include

$$110^\omega = 11, \quad 1011, \quad \text{and} \quad (10)^\omega.$$

As we can see from this example, it is possible to have multiple beta-expansions for the same number. In fact, in the case of the golden ratio, 1 has an infinite number of beta-expansions. The number of expansions of 1 has been studied in Erdős *et al.*,[22] where they showed that for all $1 \leq N \leq \omega$ there exists 2^ω real numbers $\beta \in (1, 2)$ for which 1 has exactly N different beta-expansions.

Problems relating to the structure of a general beta-expansion can be found in Joó *et al.*[30] In particular, they related the properties of these beta-expansions to particular properties of the spectrum of β defined as

$$Y(\beta) = \left\{ \sum_{i=0}^{k} \beta^{n_i} : k \in \mathbb{N}, n_i \in \mathbb{N}, n_i < n_{i+1} \right\} = \{0 = y_0 < y_1 < y_2 < \cdots \}.$$

The two specific constants $\ell(\beta) = \liminf(y_n - y_{n-1})$ and $L(\beta) = \limsup(y_n - y_{n-1})$ related to this spectrum have received much attention recently. See for example Borwein *et al.*,[9,10] Bugeaud,[18] Feng *et al.*,[24] Garth *et al.*[27] and Hare.[29]

If we have multiple beta-expansions for a value, then we can order these beta-expansions lexicographically. Of particular interest are the largest and the smallest beta-expansions lexicographically. We begin with the largest, which has received the most attention in the literature. This is called the *greedy expansion*. (The name is suggested by the fact that the algorithm used to compute this expansion is a greedy algorithm.)

Definition 1.2. If $d_\beta(x) = a_1 a_2 a_3 \cdots$ is the maximal beta-expansion for x (lexicographically) then we say that $a_1 a_2 a_3 \cdots$ is the *greedy expansion* for x with base β.

It should be noted that in some literature when they talk about the beta-expansion of x, they mean the greedy expansion of x. To avoid confusion, we will always specify an expansion as a greedy expansion or a lazy expansion if it is one of these expansions. The lazy expansion is the smallest beta-expansion lexicographically. In this paper, when we say beta-expansion we are not assuming that it is the greedy expansion, but instead that it is a general expansion of the form given in Definition 1.1.

The beta-expansions that Rényi studied were these greedy expansions. The majority of his paper was studying more general expansions such as the continued fraction expansion, or the regular q-adic expansion, where $q \in \mathbb{Z}$, with $q \geq 2$.

Rényi studied expansions of the form

$$x = \epsilon_0 + f(\epsilon_1 + f(\epsilon_2 + \cdots) \cdots)$$

with digits ϵ_n. In the case of the q-adic expansion we have $f(x) = x/q$, and in the case of the continued fraction expansion we have $f(x) = 1/x$. We can compute the digits by letting $\phi(x) = q \cdot x$ (or $\phi(x) = 1/x$ respectively) be the inverse of $f(x)$, and defining $r_n(x) = \phi(r_{n-1}(x)) \mod 1$. Then the digits are computed by $\epsilon_n = \lfloor \phi(r_{n-1}(x)) \rfloor$. Rényi showed that for an algorithm such as the continued fraction, or q-adic expansion algorithm, the digits are independent. He then considered the obvious extension of the q-adic expansion to the greedy expansion given by $f(x) = x/\beta$ with $\beta \in \mathbb{R} \setminus \mathbb{Z}$, $\beta > 1$ and $\phi(x) = \beta \cdot x \mod 1$. He observed that for the greedy expansion, the digits were not independent. An easy example of this is the case of $\beta = 1.618 \cdots$ the greater root of $x^2 - x - 1$. The greedy expansion cannot contain as a substring "011". Assume to the contrary that the greedy expansion for some x contains the substring "011". If we replace the substring "011" with "100" then we get an equally valid beta-expansion for the same number, but the latter is lexicographically bigger, contradicting the assumption that the original beta-expansion was the greedy expansion.

Algorithmically, it is straightforward to compute the greedy expansion. If the greedy expansion is eventually periodic, then it is possible to detect this, and give the complete expansion. (This requires maintaining the symbolic representation of r_n to allow for exact comparisons. See the discussion in Section 3.) If the greedy expansion is not eventually periodic, then it is possible to compute any number of terms of the expansion, relatively quickly.

Algorithm 1.1 (Greedy Algorithm). *Set* $r_0 := x$. *Set* $r_n = \beta \cdot r_{n-1}$ (mod 1) *and* $a_n = \lfloor \beta \cdot r_{n-1} \rfloor$. *Then* $a_1 a_2 a_3 \cdots$ *is the greedy expansion of* x.

Some implementation issues are discussed in Section 3. Some examples of greedy expansions are given in Table 1.

We see that one of the key computations in Rényi's study (and the greedy algorithm) is $r_n(x) = \phi(r_{n-1}(x))$ (mod 1). From this we define the operator $T_\beta(x)$ as $T_\beta(x) = \beta \cdot x$ (mod 1). We see that a necessary and sufficient condition for the greedy expansion being periodic or finite is that $T_\beta^{(n)}(x)$ is eventually periodic or eventually 0. To that end we define

Definition 1.3. We define Fin(β) as the set of all x such that $T_\beta^{(n)}(x)$ is eventually 0.

Definition 1.4. We define Per(β) as the set of all x such that $T_\beta^{(n)}(x)$ is eventually periodic.

We see that if $T_\beta^{(n)}(x)$ is eventually 0, then it is eventually periodic, with a period of length 1, and period "0". Hence Fin$(\beta) \subset$ Per(β). We notice that

Table 1. Table of some greedy and lazy expansions

x	Greedy Expansions	Lazy Expansions
	$\beta \approx 1.618$ the root of $\beta^2 - \beta - 1$	
1/5	$(0001001010101001001000)^\omega$	$000(01101^6 0110101011)^\omega$
2/5	$(010^6 100101010010)^\omega$	$0(010101101101^6 011)^\omega$
3/5	$(0101010010010^6 10)^\omega$	$0(01^6 0110101011011)^\omega$
4/5	$(10010010^6 1001010)^\omega$	$0(11011010101101101111)^\omega$
1	11	$0(1)^\omega$
	$\beta \approx 1.325$ the root of $\beta^3 - \beta - 1$	
1/5	$(0^5 10^8 10^9)^\omega$	$0^9(110111101111101^5 01^5)^\omega$
2/5	$(00010^5 1000010^9)^\omega$	$0^7(1^5 01^6 01^5 01^5)^\omega$
3/5	$(010^{10} 10^6 10000)^\omega$	$0^5 10111(111101^{12} 01^5 0)^\omega$
4/5	$1000(0^7 100001000010^5 1)^\omega$	$0000(101^8 01^{13})^\omega$
1	10001	$0000(1)^\omega$
	$\beta = 2$	
1/5	$(0011)^\omega$	$(0011)^\omega$
2/5	$(0110)^\omega$	$(0110)^\omega$
3/5	$(1001)^\omega$	$(1001)^\omega$
4/5	$(1100)^\omega$	$(1100)^\omega$
1	$(1)^\omega$	$(1)^\omega$

if β is an integer, then $\mathrm{Fin}(\beta)$ is exactly the set of numbers $\frac{n}{\beta^k}$ with $n \in \mathbb{Z}$ and $\mathrm{Per}(\beta)$ is the rationals.

One interesting difference between $\mathrm{Fin}(\beta)$ and $\mathrm{Per}(\beta)$ from when β is an integer, and when β is a general real number, is that when β is an integer, both of $\mathrm{Fin}(\beta)$ and $\mathrm{Per}(\beta)$ are necessarily closed under addition and multiplication. Moreover tight bounds upon the length of the fractional part can be given, based on the lengths of the fractional parts of the two terms to be added or multiplied. It is not necessarily true that these sets are closed under addition or multiplication for a general real number β. For some specific real numbers β with special algebraic properties, this can still be done. See for example Ambrož[6] for results and software. An obvious question occurs. What do $\mathrm{Fin}(\beta)$ and $\mathrm{Per}(\beta)$ look like in general, when β is not an integer? When can we say $\mathrm{Per}(\beta)$ contains $\mathbb{Q}$? Before discussing this, we need to introduce some standard definitions from algebraic number theory.

Definition 1.5. A number α is an *algebraic integer* if it is the root of a monic integer polynomial. There is a unique monic integer polynomial $p(x)$, called the *minimal polynomial*, for which α is a root and the degree of $p(x)$ is minimal.

Definition 1.6. If α is an algebraic integer, and $p(x)$ is its minimal polynomial, then we say that all of the other roots of $p(x)$ are the **conjugates** of α.

Definition 1.7. A **Pisot number** α is a real algebraic integer $\alpha > 1$ such that all of α's conjugates are strictly less than 1 in modulus.

Definition 1.8. A **Salem number** α is a real algebraic integer $\alpha > 1$ such that all of α's conjugates are less than or equal to 1 in modulus, and at least one conjugate is equal to 1 in modulus.

Example 1.2. An example of an algebraic integer is $1 + \sqrt{2} + \sqrt{3}$ with minimal polynomial $x^4 - 4x^3 - 4x^2 + 16x - 8$. The conjugates of $1 + \sqrt{2} + \sqrt{3}$ are $1 \pm \sqrt{2} \pm \sqrt{3}$.

An example of a Pisot number is, $1.325 \cdots$, the root of $x^3 - x - 1$. The two conjugates of this Pisot number are both of modulus $0.8689 \cdots < 1$.

An example of a Salem number is $1.7221 \cdots$, the root of $x^4 - x^3 - x^2 - x + 1$, the other conjugates have modulus $1, 1$ and $0.5807 \cdots$.

The structure of the set of all Pisot numbers is well understood. The set is known to be closed (Salem[42]), with a smallest value of $1.324 \cdots$, the real root of $x^3 - x - 1$ (Smyth[44]). Amara[3] gave a complete description of the set of all limit points of the Pisot numbers. Boyd[12,13] has given an algorithm that will find all Pisot numbers in an interval, where, in the case of limit points, the algorithm can detect the limit points and compensate for them. Both of these results were exploited for the computational exploration of beta-expansions of Pisot numbers given in Allouche *et al.*,[2] (see Section 2 for more details).

The set of Salem numbers is not as well understood. It is known that every Pisot number is the limit of Salem numbers from both above and below. No smallest value in the set of Salem numbers is known. In fact, a major open conjecture is if there is a smallest Salem number, and if so, what is it? See for example Borwein[8] for more discussion of this and related conjectures. The smallest known Salem number is $1.1762 \cdots$, the root of $x^{10} + x^9 - x^7 - x^6 - x^5 - x^4 - x^3 + x + 1$, which was found in 1933 by Lehmer.[37] Despite numerous computer searches since then, no better example has been found. See for example Boyd[11,14] and Mossinghoff.[39]

So, now that we have some of the basic tools, we can talk about what $\text{Fin}(\beta)$ and $\text{Per}(\beta)$ look like when β is a non-integer. In Bertrand,[7] it is shown that if β is a Pisot number then a necessary and sufficient condition for x to have an eventually periodic beta-expansion, is that $x \in \mathbb{Q}(\beta)$. Schmidt[43] showed that $\mathbb{Q} \cap [0,1) \subset \text{Per}(\beta)$ implies that β is a Pisot or a Salem number. Moreover, he showed that if β is not a Pisot nor a Salem number, then $\text{Per}(\beta) \cap \mathbb{Q}$ is nowhere dense in $[0,1)$. An important conjecture, that has motivated a lot of later work is that

Schmidt's Conjecture. *We have $\mathbb{Q} \cap [0,1) \subset \text{Per}(\beta)$ if and only if β is a Pisot or a Salem number.*

One direction is known already, namely that $[0,1) \cap \mathbb{Q} \subset \text{Per}(\beta) \cup \text{Fin}(\beta)$ implies that β is a Salem or Pisot number. The other direction is shown only for the Pisot number case. All that remains to show is that β being a Salem number implies that $\mathbb{Q} \cap [0,1) \subset \text{Per}(\beta)$.

A related question to this, first investigated by Parry,[40] concerns when $1 \in \text{Per}(\beta)$. Another way to phrasing this is determining for which β do we have $d_\beta(1)$ is either eventually periodic or finite. We say that β is a beta-number if $d_\beta(1)$ is eventually periodic, and we say that β is a simple beta-number if $d_\beta(1)$ is finite. Much like the situation with Schmidt's conjecture, it is known that if β is a Pisot number, then β is a beta-number. A very simple argument of Boyd's[14] shows that if β is a Salem number, then $d_\beta(1)$ cannot be finite, and hence cannot be a simple beta-number. But this raised the question, how often can Salem numbers be beta-numbers?

Motivated by this Boyd looks at the greedy expansion of 1. Boyd[14] shows that if β is a Salem number of degree 4 then $1 \in \text{Per}(\beta)$. In fact, he proved if β is a degree 4 Salem number, then $d_\beta(1)$ had a pre-period of length 1 and a period of length 3, 5, 9 or an even number. Furthermore, the length of the period is bounded by $2\beta + 3$. He gave heuristical evidence to show that if β is a Salem number then β would be a beta-number for degree six, but for higher degrees, β would not be a beta-number a positive proposition of the time.

Consider the expansion $d_\beta(1) = a_1 a_2 \cdots a_k$ if the greedy expansion is finite, and $d_\beta(1) = a_1 a_2 \cdots a_k (a_{k+1} \cdots a_n)^\omega$ if the greedy expansion is eventually periodic. Define $P_i(x) = x^i - a_1 x^{i-1} - \cdots - a_i$. We define the companion polynomial as:

$$R(x) = \begin{cases} P_n(x) & \text{if } d_\beta(1) \text{ is finite} \\ P_n(x) - P_k(x) & \text{if } d_\beta(1) \text{ is eventually periodic} \end{cases}$$

By considering β as a beta-number (or simple beta-number), we see that $R(x)$ is a well defined monic integer polynomial. Moreover $R(\beta) = 0$. Thus we see that β must be an algebraic integer. Let $p(x)$ be the minimal polynomial for β, then we see that $p(x)|R(x)$. In fact we can write

$R(x) = p(x)Q(x)$. Here $Q(x)$ is called the co-factor of the beta-expansion, where as $R(x)$ is called the companion polynomial.

An equivalent analysis can be done for $d_\beta(\alpha)$ by defining $P_i(x) = \alpha x^i - a_1 x^{i-1} - \cdots - a_i$. To complete this analysis we would have to look at the algebraic properties of α as well as β. This means we may not be able to restrict our attention to algebraic integers. For the purposes of this paper, we assume that we are considering the greedy expansion of 1, and hence $\alpha = 1$.

The study of the companion polynomial, such as the location of its roots was initiated by Parry. In particular Parry[40] showed the roots of $R(x)$, other than β are in $|z| < \min(2, \beta)$. This was improved to $|z| \le \frac{1+\sqrt{5}}{2}$ by Solomyak[45] and independently by Flatto et al.[25]

Boyd proved that the co-factor is reciprocal and cyclotomic when β is a degree 4 Salem number. Boyd[15,17] also gave heuristics to show that this is true for degree 4 and 6, but not true for a positive proportion of degree 8 or more. Boyd[15] did a massive computation to check the degree 6 case. He checked the co-factors of 11836 degree 6 Salem numbers (of reasonably small trace).

Based on preliminary experimental evidence, it was conjectured that the complementary factor was always cyclotomic (and hence reciprocal). Boyd[15] showed that this is false for Pisot numbers, by doing a large search over a particular set of Pisot numbers. In particular, he looked at:

- All Pisot numbers up to degree 50 in $[1.9, 2]$
- All Pisot numbers up to degree 60 in $[1.96, 2]$
- All Pisot numbers up to degree 20 in $[2, 2.2]$
- All irregular Pisot numbers in $[1, 1.9324] \cup [1.9333, 1.96]$
- Pisot numbers associated with the first 8 limit points.

(See Theorem 2.1 and Table 2 for a discussion of regular and irregular Pisot numbers, as well as for the structure of the limit points of Pisot numbers.) Boyd found examples of Pisot numbers whose co-factors were both non-reciprocal as well as reciprocal but non-cyclotomic. He also found infinite families of examples of each of these.

Most of our discussion so far has been derived from properties of the greedy expansion (the maximal expansion lexicographically.) At this point, we need to introduce the other important beta-expansion. Namely, the minimal beta-expansion lexicographically, the *lazy expansion*.

Definition 1.9. If $\ell_\beta(x) = a_1 a_2 a_3 \cdots$ is the minimal beta-expansion for x (lexicographically) then we say that $a_1 a_2 a_3 \cdots$ is a *lazy expansion*.

Algorithmically, the lazy expansion is easy to compute, with the same caveat as before that if the expansion is not eventually periodic or finite, then the best we can hope for is to be able to compute a large number of terms.

Algorithm 1.2 (Lazy Algorithm). *Consider a sequence*

$$\frac{a_1}{\beta} + \frac{a_2}{\beta^2} + \cdots + \frac{a_k}{\beta^k} + \frac{?}{\beta^{k+1}} + \cdots$$

where the $a_1, \cdots, a_k$ have previously been determined, and we wish to determine a_{k+1}. We choose $a_{k+1} \in \{0, 1, \cdots, \lceil \beta \rceil - 1\}$, minimal such that

$$\frac{a_1}{\beta} + \frac{a_2}{\beta^2} + \cdots + \frac{a_k}{\beta^k} + \frac{a_{k+1}}{\beta^{k+1}} + \frac{\lceil \beta \rceil - 1}{\beta^{k+2}} + \frac{\lceil \beta \rceil - 1}{\beta^{k+3}} + \frac{\lceil \beta \rceil - 1}{\beta^{k+4}} + \cdots \geq \alpha.$$

The expansion is denoted by $\ell_\beta(\alpha)$.

Implementation issues are discussed in Section 3. Some examples of lazy expansions are given in Table 1.

There has been a study of numbers β such that the greedy and lazy expansions of 1 are equal. In Erdos *et al.*,[21] a combinatorial method of determining when an expansion is greedy or unique is given. We call a number β *univoque* if $d_\beta(1) = \ell_\beta(1)$ (or equivalently, if 1 has a unique expansion base β). The set of all such numbers is defined by $\mathcal{U}$. Of the set of univoque numbers, $\mathcal{U}$, there is a smallest such number (Komornik *et al.*[33]), $\kappa \approx 1.787231 \cdots$. Moreover, κ is transcendental (Allouche *et al.*[1]) and is not isolated (Komornik *et al.*[36]). Komornik *et al.*[35] showed that $\overline{\mathcal{U}}$ is a perfect set, and moreover, that since $\mathcal{U}$ has measure 0, then $\mathcal{U}$ is a Cantor set. In the case when the expansions are periodic or finite, this method can be made algorithmic (Allouche *et al.*[2]). In particular, in Allouche *et al.*[2] the authors look at Pisot numbers with respect to these expansions. This is explained in more detail in Section 2.

When the greedy and lazy expansions are not equal, occasionally there are an infinite number of different general beta-expansions. In Dajani *et al.*[19] and de Vries *et al.*[20] it is shown how to create "quasi-greedy" expansions, which will fall in between the greedy and the lazy expansions.

Another, even more general idea is given in Komornik *et al.*,[34] where they generalized the idea of expansions to $\sum c_i \cdot p_i$ where $c_i \in \mathbb{Z}$, $0 \leq c_i \leq m_i$ and $p_i \to 0$. This is the same as the standard beta-expansion when $p_i = \beta^{-i}$ and $m_i = \lceil \beta \rceil$. Another interesting variation on the idea of beta-expansions includes the situation where the base β is a complex number. For example, see Chapter 7 or Lothaire,[38] Ambrož *et al.*,[5] and Komornik *et al.*[32]

2. Univoque Pisot Numbers

The goal of the talk at WWCA was to discuss some recent work of the author with Allouche and Frougny concerning their investigation of univoque Pisot numbers. Recall a Pisot number is a real root greater than 1 of a monic integer polynomial, such that all of its conjugates have modulus strictly less than 1. A number β is univoque if the greedy and lazy expansions of 1 are the same. The main questions of the investigation were:

- Are there any univoque Pisot numbers?
- Is there a smallest univoque Pisot number?
- Are there any infinite families of univoque Pisot numbers?
- What sort of structure does the set of univoque Pisot numbers have?

The answer to the first three questions were all yes. Before discussing the last question (and how we arrived at the first three answers) we need to discuss the structure of the set of Pisot numbers in more detail.

We denote the set of Pisot numbers by S. Amara[3] has determined all the limit points of S smaller than 2.

Theorem 2.1. *The limit points of S in $(1, 2)$ are the following:*

$$\varphi_1 = \psi_1 < \varphi_2 < \psi_2 < \varphi_3 < \chi < \psi_3 < \varphi_4 < \cdots < \psi_r < \varphi_{r+1} < \cdots < 2$$

where

$$\begin{cases} \text{the minimal polynomial of } \varphi_r \text{ is } \Phi_r(x) = x^{r+1} - 2x^r + x - 1, \\ \text{the minimal polynomial of } \psi_r \text{ is } \Psi_r(x) = x^{r+1} - x^r - \cdots - x - 1, \\ \text{the minimal polynomial of } \chi \text{ is } \mathcal{X}(x) = x^4 - x^3 - 2x^2 + 1. \end{cases}$$

The first few limit points are:

- $\varphi_1 = \psi_1 \approx 1.618033989$, the root in $(1, 2)$ of $\Psi_1(x) = \Phi_1(x) = x^2 - x - 1$
- $\varphi_2 \approx 1.754877666$, the root in $(1, 2)$ of $\Psi_2(x) = x^3 - 2x^2 + x - 1$
- $\psi_2 \approx 1.839286755$, the root in $(1, 2)$ of $\Psi_2(x) = x^3 - x^2 - x - 1$
- $\varphi_3 \approx 1.866760399$, the root in $(1, 2)$ of $\Phi_3(x) = x^4 - 2x^3 + x - 1$
- $\chi \approx 1.905166168$, the root in $(1, 2)$ of $\mathcal{X}(x) = x^4 - x^3 - 2x^2 + 1$
- $\psi_3 \approx 1.927561975$, the root in $(1, 2)$ of $\Psi_3(x) = x^4 - x^3 - x^2 - x - 1$

A description of the Pisot numbers approaching these limit points was given by Talmoudi.[46] Regular Pisot numbers are defined as the Pisot roots of the polynomials in Table 2. Pisot numbers that are not regular Pisot numbers are called irregular Pisot numbers. For each of these limit points (φ_r, ψ_r or χ), there exists an ϵ, (dependent on the limit point) such that all Pisot

Table 2. Regular Pisot numbers

Limit Points	Defining polynomials
φ_r	$\Phi_r(x)x^n \pm (x^r - x^{r-1} + 1)$
	$\Phi_r(x)x^n \pm (x^r - x + 1)$
	$\Phi_r(x)x^n \pm (x^r + 1)(x - 1)$
ψ_r	$\Psi_r(x)x^n \pm (x^{r+1} - 1)$
	$\Psi_r(x)x^n \pm (x^r - 1)/(x - 1)$
χ	$\mathcal{X}(x)x^n \pm (x^3 + x^2 - x - 1)$
	$\mathcal{X}(x)x^n \pm (x^4 - x^2 + 1)$

numbers in an ϵ-neighbourhood of this limit point are these regular Pisot numbers. The Pisot root of the defining polynomial approaches the limit point as n tends to infinity. The defining polynomials for these regular Pisot numbers are given in Table 2. It should be noted that these polynomials are not necessarily minimal, and may contain some cyclotomic factors. Also, they are only guaranteed to have a Pisot number root for sufficiently large n (although for our purposes, $n = 1$ or 2 normally is sufficiently large).

The key observation used for this study was that "nice" sequences of regular Pisot numbers approaching a limit point give "nice" sequences of greedy and lazy expansions. This meant that it was possible to determine infinite classes of Pisot numbers as being univoque or not univoque. Along with the observation that χ was a univoque Pisot number, our new goal became:

- Find the greedy and lazy expansions of all regular Pisot numbers approaching limit points less than or equal to χ.
- Find all Pisot numbers less than χ by Boyd's algorithm, removing from the search those regular Pisot numbers accounted for in the previous step. After this, determine which of these Pisot numbers are univoque.

Table 3. Greedy and lazy β-expansions of real numbers in $S' \cap (1, 2)$.

Minimal Polynomial	Pisot Number	Greedy expansion	Lazy expansion	Comment
$x^{r+1} - 2x^r + x - 1$	φ_r	$1^r 0^{r-1} 1$	$1^{r-1} 01^\omega$	
$x^{r+1} - x^r - \cdots - 1$	ψ_r	1^{r+1}	$(1^r 0)^\omega$	
$x^4 - x^3 - 2x^2 + 1$	χ	$11(10)^\omega$	$11(10)^\omega$	Univoque

Some examples of the types of patterns found are given in Table 3 and 4. For a more complete list, see Allouche *et al.*[2] Table 3 shows us that χ is a univoque Pisot number. Moreover, we see from Table 4 that χ is the limit point of univoque Pisot numbers. Using a similar analysis, it was shown

Table 4.   Greedy and lazy expansions for some regular Pisot numbers.

Case	Greedy expansion	Lazy expansion	Comment
	$\Psi_2(x)x^n - (x+1)$		
$n = 1$	Root bigger than 2		
$n = 4$	1110011	$(1110010)^\omega$	
$n = 3k+1$	$11100(000)^{k-1}11$	$(11(011)^{k-1}1001(101)^{k-1}0)^\omega$	
	$\mathcal{X}(x)x^n - (x^3 + x^2 - x - 1)$		
$n = 2$	Root bigger than 2		
$n = 2k+2$	$111(01)^{k-1}1011((10)^{k-1}0111(01)^{k-1}1000)^\omega$		Univoque
		$111(01)^{k-1}1011((10)^{k-1}0111(01)^{k-1}1000)^\omega$	

that there were no other regular Pisot numbers approaching a limit point less that χ that was univoque. Using Boyd's method,[12,13] all relevant Pisot numbers, not accounted for above, less than χ were enumerated and tested to see if they were univoque. (Some intervals of Pisot numbers could be eliminated based on combinatorial arguments that are not of interest to this survey.) This gives us the result that there are exactly two univoque Pisot numbers less than χ. They are

- $1.880000\cdots$ the root in $(1,2)$ of the polynomial $x^{14} - 2x^{13} + x^{11} - x^{10} - x^7 + x^6 - x^4 + x^3 - x + 1$ with univoque expansion $111001011(1001010)^\omega$.
- $1.886681\cdots$ the root in $(1,2)$ of the polynomial $x^{12} - 2x^{11} + x^{10} - 2x^9 + x^8 - x^3 + x^2 - x + 1$ with univoque expansion $111001101(1100)^\omega$

3. Algorithms and Implementation Issues

One of the main computations needed for computing the greedy expansion is the calculation of $a_n = \lfloor \beta \cdot r_{n-1} \rfloor$. This must be done as a floating point calculation, as Maple is unable to determine which integer this should be symbolically. Unfortunately the introduction of floating point numbers allows for the introduction of rounding error. To protect against rounding error, we test $|a_n - \beta \cdot r_{n-1}|$ to a tolerance *tol*. For our purposes, we use *tol* as $10^{-Digits/2}$ where *Digits* is the number of digits of accuracy we are computing to. If we have that $|a_n - \beta \cdot r_n| < tol$ then we assume that we are not working with a high enough accuracy, and we double *Digits*, recompute β to this higher degree of accuracy, and then redo the calculation.

In contrast to this, the calculation for r_n is always done symbolically. This is done for two reasons. The first is that it prevents the accumulation of round-off error through successive calculations. The second reason is that it allows for the detection of the beta-expansion begin periodic or finite, as we compare r_n to all r_k with $k < n$. Similar issues needed to be addressed for computing the lazy expansion. An example of the code being run is given below. The code itself can be found at author's webpage.[28]

Example 3.1. The code below assumes that β is the unique root of the polynomial between 1 and 2. Also, as Pisot numbers are known to be beta-numbers, they are always completely determined. When the polynomial does not have a Pisot root, then there is no guarantee that the root is a beta-number, so the greedy or lazy expansion may not have a nice closed form. For this reason, only the first 60 digits are computed. If in the first 60 digits, the expansion is determined to be periodic, then this information is presented, otherwise, only the first 60 digits are presented. By default, the expansion of 1 is given.

```
> read all;
> PPGreedy(x^3-x-1);
                        "10001(0)^omega"

> PPLazy(x^3-x-1);
                        "0000(1)^omega"

> PPGreedy(x^4-x-1);
"Warning: Polynomial is not Pisot.  Only Computing first 60 terms"
    "100000001000000000000001000000000100000000000100000000000000000001..."

> PPLazy(x^4-x-1);
"Warning: Polynomial is not Pisot.  Only Computing first 60 terms"
    "000000011101111111111111101111111110111111111110111111111111111111..."
```

One interesting observation made, which allows for the classification of regular Pisot numbers is that "nice" sequences of regular Pisot numbers give "nice" sequences of beta expansions.

Example 3.2.

```
> read all;
> for n from 1 to 5 do n, 'PPGreedy'(x^(n+1)-2*x^n+x-1); od;
                    1, "11(0)^omega"

                    2, "1101(0)^omega"

                    3, "111001(0)^omega"

                    4, "11110001(0)^omega"
```

```
                    5, "1111100001(0)^omega"

> for n from 1 to 5 do n, 'PPLazy'(x^(n+1)-2*x^n+x-1); od;
                    1, "0(1)^omega"

                    2, "10(1)^omega"

                    3, "110(1)^omega"

                    4, "1110(1)^omega"

                    5, "11110(1)^omega"
```

Now some sequences of expansions are "nicer" than others, and it would be time consuming and prone to errors to find all of these patterns by hand. For that reason, algorithms were developed that:

- Predicted the sequences of expansions based upon the first few terms in the sequence.
- Based upon this prediction, predicted the sequence of companion polynomials.
- Based upon these companion polynomials, showed that the defining polynomial divides the companion polynomial, thus showing that the predicted sequences gives a valid sequence of expansions. This actually required the prediction of the co-factor polynomials.
- Use combinatorial properties (not discussed here) to show that these predicted sequences of expansions are greedy or lazy expansions as necessary, (see Allouche *et al.*[2] for more details).

Example 3.3. In the next example, we first have a strange definition of P, as every other defining polynomial has a cyclotomic factor of $x + 1$. When we list six terms, no patterns becomes apparent for the lazy expansion. But if we look at the every third expansion, a pattern does present itself, and it is easy to prove this pattern.

```
> read all;
> P := (n,x)->'if'(type(n, odd),
        (x^3-x^2-x-1)*x^n-x^3+1,
        simplify(((x^3-x^2-x-1)*x^n-x^3+1)/(x+1))):

> for n from 3 to 8 do n, PPGreedy(P(n,x)), PPLazy(P(n,x));od;
          3, "111(110)^omega", "111(110)^omega"
```

```
        4, "111(0110)^omega", "111(0110)^omega"

     5, "111(00110)^omega", "1110010111(11110)^omega"

      6, "111(000110)^omega", "110111(110)^omega"

      7, "111(0000110)^omega", "110111(0110110)^omega"

  8, "111(00000110)^omega", "1101110011010111(11011110)^omega"

> for n from 3 to 18 by 3  do n, PPGreedy(P(n,x)), PPLazy(P(n,x));od;
        3, "111(110)^omega", "111(110)^omega"

      6, "111(000110)^omega", "110111(110)^omega"

      9, "111(000000110)^omega", "110110111(110)^omega"

  12, "111(000000000110)^omega", "110110110111(110)^omega"

  15, "111(000000000000110)^omega", "110110110110111(110)^omega"

 18, "111(000000000000000110)^omega", "110110110110110111(110)^omega"

> CompleteConj2(n->P(n, x), 6, 3, k, x);
Looking at n = 3 k + 3
Greedy Expansion: 111((000)^(k) 110)^omega
- Univoque Exceptions: k = 0 is univoque
- Greedy Expansion valid for k >= 0
Lazy Expansion: 11(011)^(k) 1(110)^omega
- Lazy Expansion is valid for k >= 0
```

The way the output from CompleteConj2 should be interpreted is

- Let β_n be the Pisot root of $(x^3 - x^2 - x - 1)x^n - x^3 + 1$.
- Then the greedy expansion for β_{3k+3} is $111((000)^k 110)^\omega$.
- The lazy expansion for β_{3k+3} is $11(011)^k 1(110)^\omega$.
- Both of these expansions are valid for all k.
- In the case $k = 0$ this expansion is univoque.

4. Conclusions and Open Questions

For convenience we restate the three conjectures that were given in this paper.

Schmidt's Conjecture. *We have $\mathbb{Q} \cap [0,1) \subset \mathrm{Per}(\beta)$ if and only if β is a Pisot or a Salem number.*

Boyd's Conjecture. *The greedy expansion $d_\beta(1)$ is eventually periodic for Salem numbers β of degree 6, but the expansion is not eventually periodic for a positive proportion of Salem numbers of higher degrees.*

Lehmer's Conjecture. *The smallest Salem number is $1.1762\cdots$, the root of $x^{10} + x^9 - x^7 - x^6 - x^5 - x^4 - x^3 + x + 1$.*

Some other interesting questions worth investigating are:

- In general, are the greedy/lazy β-expansions periodic for Salem numbers? (This is not known to be true, see Boyd[16] for more details.)
- It is known that Pisot numbers can be written as a limit of Salem numbers, where if $P(x)$ is the minimal polynomial of a Pisot number, then $P(x)x^n \pm P^*(x)$ has a Salem number as a root, which tends to the root of the Pisot number. Some preliminary and somewhat haphazard investigation suggests that we might be able to find a "nice" looking expression for the greedy (resp. lazy) beta-expansion of these Salem numbers, which tends towards the greedy (resp. lazy) beta-expansion of the Pisot number. If true, then this could have implications towards questions concerning the beta-expansions of Salem numbers being eventually periodic.

References

1. J.-P. Allouche and M. Cosnard. The Komornik-Loreti constant is transcendental. *Amer. Math. Monthly*, 107(5):448–449, 2000.
2. J.-P. Allouche, C. Frougny, and K. G. Hare. On Univoque Pisot numbers. *Math. Comp.*, to appear.
3. M. Amara. Ensembles fermés de nombres algébriques. *Ann. Sci. École Norm. Sup. (3)*, 83:215–270 (1967), 1966.
4. P. Ambrož. On the tau-adic expansions of real numbers. In S. Brlek and C. Reutenauer, editors, *Words 2005, 5th International Conference on Words, actes*, volume 36 of *Publications du LaCIM*, pages 79–89. UQÀM, 2005.
5. P. Ambrož and Christiane Frougny. On alpha-adic expansion of Pisot numbers. Submitted to Theoret. Comp. Sci., 2006.
6. P. Ambrož. Pisot numbers and numeration systems. Master's thesis, FNSPE Czech Technical University, 2003. Software and documentation available from World Wide Web (http://linux.fjfi.cvut.cz/~ampy/).
7. A. Bertrand. Développements en base de Pisot et répartition modulo 1. *C. R. Acad. Sci. Paris Sér. A-B*, 285(6):A419–A421, 1977.
8. P. Borwein. *Computational excursions in analysis and number theory*, volume 10 of *CMS Books in Mathematics/Ouvrages de Mathématiques de la SMC*. Springer-Verlag, New York, 2002.
9. P. Borwein and K. G. Hare. Some computations on the spectra of Pisot and Salem numbers. *Math. Comp.*, 71(238):767–780 (electronic), 2002.

10. P. Borwein and K. G. Hare. General forms for minimal spectral values for a class of quadratic Pisot numbers. *Bull. London Math. Soc.*, 35(1):47–54, 2003.
11. D. W. Boyd. Reciprocal polynomials having small measure. *Math. Comp.*, 35(152):1361–1377, 1980.
12. D. W. Boyd. Pisot numbers in the neighborhood of a limit point. II. *Math. Comp.*, 43(168):593–602, 1984.
13. D. W. Boyd. Pisot numbers in the neighbourhood of a limit point. I. *J. Number Theory*, 21(1):17–43, 1985.
14. D. W. Boyd. Reciprocal polynomials having small measure. II. *Math. Comp.*, 53(187):355–357, S1–S5, 1989.
15. D. W. Boyd. On beta expansions for Pisot numbers. *Math. Comp.*, 65(214):841–860, 1996.
16. D. W. Boyd. On the beta expansion for Salem numbers of degree 6. *Math. Comp.*, 65(214):861–875, S29–S31, 1996.
17. D. W. Boyd. The beta expansion for Salem numbers. In *Organic mathematics (Burnaby, BC, 1995)*, volume 20 of *CMS Conf. Proc.*, pages 117–131. Amer. Math. Soc., Providence, RI, 1997.
18. Y. Bugeaud. Sur la suite des nombres de la forme $q^{n_1} + \cdots + q^{n_k}$. *Arch. Math. (Basel)*, 79(1):34–38, 2002.
19. K. Dajani and C. Kraaikamp. From greedy to lazy expansions and their driving dynamics. *Expo. Math.*, 20(4):315–327, 2002.
20. M. de Vries. *Random β-expansions, unique expansions and Lochs' Theorem.* PhD thesis, Vrije Universiteit, Amsterdam, 2005.
21. P. Erdős, I. Joó, and V. Komornik. Characterization of the unique expansions $1 = \sum_{i=1}^{\infty} q^{-n_i}$ and related problems. *Bull. Soc. Math. France*, 118(3):377–390, 1990.
22. P. Erdős, I. Joó, and V. Komornik. On the number of q-expansions. *Ann. Univ. Sci. Budapest. Eötvös Sect. Math.*, 37:109–118, 1994.
23. P. Erdős and V. Komornik. Developments in non-integer bases. *Acta Math. Hungar.*, 79(1-2):57–83, 1998.
24. D.-J. Feng and Z.-Y. Wen. A property of Pisot numbers. *J. Number Theory*, 97(2):305–316, 2002.
25. L. Flatto, J. C. Lagarias, and B. Poonen. The zeta function of the beta transformation. *Ergodic Theory Dynam. Systems*, 14(2):237–266, 1994.
26. C. Frougny and B. Solomyak. Finite beta-expansions. *Ergodic Theory Dynam. Systems*, 12(4):713–723, 1992.
27. D. Garth and K. G. Hare. Comments on the spectra of Pisot numbers. *J. Number Theory*, **121**, 187 (2006).
28. K. G. Hare. Home page. http://www.math.uwaterloo.ca/~kghare, 2004.
29. K. G. Hare. The structure of the spectra of Pisot numbers. *J. Number Theory*, 105(2):262–274, 2004.
30. I. Joó and F. J. Schnitzer. On some problems concerning expansions by noninteger bases. *Anz. Österreich. Akad. Wiss. Math.-Natur. Kl.*, 133:3–10 (1997), 1996.
31. V. J. Katz. *A history of mathematics*. HarperCollins College Publishers, New York, 1993. An introduction.
32. V. Komornik and P. Loreti. Expansions in complex bases. *Canad. Math. Bull.*

(to appear).

33. V. Komornik and P. Loreti. Unique developments in non-integer bases. *Amer. Math. Monthly*, 105(7):636–639, 1998.

34. V. Komornik and P. Loreti. Subexpansions, superexpansions and uniqueness properties in non-integer bases. *Period. Math. Hungar.*, 44(2):197–218, 2002.

35. V. Komornik and P. Loreti. On the topological structure of univoque sets. *J. Number Theory*, **122**, 1 (2007).

36. V. Komornik, P. Loreti, and A. Pethő. The smallest univoque number is not isolated. *Publ. Math. Debrecen*, 62(3-4):429–435, 2003. Dedicated to Professor Lajos Tamássy on the occasion of his 80th birthday.

37. D. H. Lehmer. Factorization of certain cyclotomic functions. *Ann. of Math. (2)*, 34(3):461–479, 1933.

38. M. Lothaire. *Algebraic combinatorics on words*, volume 90 of *Encyclopedia of Mathematics and its Applications*. Cambridge Univ. Press, Cambridge, 2002.

39. M. J. Mossinghoff. Polynomials with small Mahler measure. *Math. Comp.*, 67(224):1697–1705, S11–S14, 1998.

40. W. Parry. On the β-expansions of real numbers. *Acta Math. Acad. Sci. Hungar.*, 11:401–416, 1960.

41. A. Rényi. Representations for real numbers and their ergodic properties. *Acta Math. Acad. Sci. Hungar*, 8:477–493, 1957.

42. R. Salem. *Algebraic numbers and Fourier analysis*. D. C. Heath and Co., Boston, Mass., 1963.

43. W. M. Schmidt. *Diophantine approximation*, volume 785 of *Lecture Notes in Mathematics*. Springer, Berlin, 1980.

44. C. J. Smyth. On the product of the conjugates outside the unit circle of an algebraic integer. *Bull. London Math. Soc.*, 3:169–175, 1971.

45. B. Solomyak. Conjugates of beta-numbers and the zero-free domain for a class of analytic functions. *Proc. London Math. Soc. (3)*, 68(3):477–498, 1994.

46. F. L. Talmoudi. Sur les nombres de $S \cap [1,2[$. *C. R. Acad. Sci. Paris Sér. A-B*, 287(10):A739–A741, 1978.

LOGARITHMIC FUNCTIONAL AND THE WEIL
RECIPROCITY LAW

KHOVANSKII A.

Department of Mathematics, University of Toronto,
Toronto, Ontario, Canada
E-mail: askold@math.toronto.edu
http://www.math.toronto.edu

Dedicated to my old friend Sergei Abramov

This article is devoted to the (one-dimensional) logarithmic functional with argument being one-dimensional cycle in the group $(\mathbb{C}^*)^2$. This functional generalizes usual logarithm, that can be viewed as zero-dimensional logarithmic functional. Logarithmic functional inherits multiplicative property of the logarithm. It generalizes the functional introduced by Beilinson for topological proof of the Weil reciprocity law. Beilinson's proof is discussed in details in this article.

Logarithmic functional can be easily generalized for multidimensional case. It's multidimensional analog (see Ref. 6) proves multidimensional reciprocity laws of Parshin. I plan to return to this topic in upcoming publications.

1. Introduction

We start introduction with a brief discussion of classic results related to this paper: reciprocity law, its topological proof and multidimensional generalization. Further we comment on the logarithmic functional and its relation to the Beilinson functional. We end introduction with layout of material.

1.1. *The Weil reciprocity law*

Given two polynomials of degrees n and m with leading coefficients equal to one, the following equality holds: the product of values of the first polynomial over the roots of the second one is equal to the product of values of the second polynomial over the roots of the first one, multiplied by $(-1)^{mn}$. André Weil has found further generalization of this equality called the *reciprocity law*. The Weil reciprocity law applies to any pair of non-zero rational functions f, g on arbitrary irreducible algebraic curve X over an algebraically closed field K. In this paper we will consider only the case of the field K being the field of complex numbers $\mathbb{C}$. We will give exact statement of the reciprocity law for $K = \mathbb{C}$ in Section 2. Now we just give general comments.

The law is as follows. For each point $a \in X$ some non-zero element $[f, g]_a$ of the field K is defined, that is called the *Weil symbol* of functions f, g at the point a. The Weil symbol depends on functions f, g skew-symmetrically, i.e. $[f, g]_a = [g, f]_a^{-1}$. Besides the Weil symbol is multiplicative in each argument, i.e. for any triplet of non-zero rational functions f, g, φ and any point $a \in X$ the following equalities hold $[f\varphi, g]_a = [f, g]_a[\varphi, g]_a$ and $[f, \varphi g]_a = [f, g]_a[f, \varphi]_a$. Let O be the union of supports of divisors of functions f and g. Weil symbol of functions f, g can differ form 1 only at points a of the finite set O. Therefore the product of Weil symbols over all points a of curve X is well defined: it is equal to the product of symbols $[f, g]_a$ over finite set of points $a \in O$. The Weil reciprocity law states that *the product of Weil symbols $[f, g]_a$ over all point of the curve X is equal to one*.

1.2. *Topological explanation of the reciprocity law over the field* $\mathbb{C}$

In the case of $K = \mathbb{C}$ Beilinson[1] proved the Weil reciprocity law topologically and has found topological generalization of the Weil symbols. Similar construction was independently obtained by Pierre Deligne, but has been published much later.[2] Let f, g be non-zero rational functions on connected complex algebraic curve X and let $O \subset X$ be finite set containing support of divisors of functions f and g. Consider piecewise-smooth oriented closed curve $\gamma : [0, 1] \to X \setminus O$, $\gamma(0) = \gamma(1)$ on the manifold X that does not contain points of set O. Beilnson introduced a functional that associates a pair f, g and curve γ with an element $B_\gamma(f, g)$ of the group $\mathbb{C}/\mathbb{Z}$ (i.e. some complex number defined up to an integer additive term). He showed that:

1) For fixed f, g the functional $B_\gamma(f, g)$ considered as a function of cycle γ gives an element of one-dimensional cohomology group of the manifold $X \setminus O$ with coefficients in group $\mathbb{C}/\mathbb{Z}$, i.e. for homologous to zero in $X \setminus O$ integer-valued linear combination $\sum k_i \gamma_i$ of oriented curves γ_i the equality $\sum k_i B_\gamma(f, g) = 0$ holds.

2) The functional skew-symmetrically depends on f and g, i.e. $B_\gamma(f, g) = -B_\gamma(g, f)$.

3) The functional has the following property of multiplicativity: for any three rational functions f, g, φ not having zeros and poles in the set $X \setminus O$ and for any curve $\gamma \subset X \setminus O$, equalities $B_\gamma(f\varphi, g) = B_\gamma(f, g) + B_\gamma(\varphi, g)$

and $B_\gamma(f, \varphi g) = B_\gamma(f, g) + B_\gamma(f, \varphi)$ hold.

4) The functional $B_\gamma(f, g)$ is related to Weil symbols as follows: for any point $a \in O$ and small curve $\gamma \subset X \setminus O$ "running around point a", equality $[f, g]_a = \exp 2\pi i B_\gamma(f, g)$ holds.

In accordance with 4) Weil symbols correspond to the values of Beilinson functional on special curves related to the points of the set O. The sum of those curves over all points in O is homologous to zero in $X \setminus O$, which gives the Weil reciprocity law (see Section 5).

Summarizing, Beilinson functional explains the Weil reciprocity law over field $\mathbb{C}$, and also provides an analog of Weil symbols for arbitrary cycles in $X \setminus O$, not necessary related to the points of the set O.

1.3. *Multi-dimensional reciprocity laws*

A.N. Parshin (Refs. 3,8,9) has found remarkable multi-dimensional generalization of the reciprocity law. It applies to an arbitrary collection $\mathbf{f} = \{f_1, \ldots, f_{n+1}\}$ of $n + 1$ nonzero rational functions f_j on irreducible n-dimensional manifold M over arbitrary algebraically closed field K. In multi-dimensional generalization instead of point on manifold M one considers flag $F = \{M_0 \subset M_1 \subset \cdots \subset M_n = M\}$ consisting of a chain of embedded germs M_j of algebraic manifolds of increasing dimensions, $\dim M_j = j$, locally irreducible in the neighborhood of the point M_0. Each collection of functions $\mathbf{f}$ and flag F is associated with nonzero element $[\mathbf{f}]_F$ of the field K, which is called *Parshin symbol*. *Reciprocity laws* state that for some chosen (precisely described) finite sets of flags L the product $\prod\limits_{F \in L} [\mathbf{f}]_F$ is equal to 1.

Let $n = 1$ and manifold M be nonsingular algebraic curve. In this case:

1) flag $F = \{M_0 \subset M\}$ is defined by point $a = M_0$,

2) Parshin symbol $[\mathbf{f}]_F$ of the pair of functions $\mathbf{f} = f_1, f_2$ on flag F coincides with Weil symbol $[f_1, f_2]_a$,

3) the only chosen set of flags L is equal to the union of supports of the divisor of functions f_1, f_2,

4) reciprocity law for this set coincides with the Weil reciprocity law.

In the case of $K = \mathbb{C}$ Brylinski and McLaughlin[7] proved multidimensional multiplicity laws topologically and found topological generalization of Parshin symbols. Their topological construction heavily uses sheaf theory and is not intuitive.

About 10 years ago I have found[5] explicit formula for the product of roots of a system of algebraic equations with general enough set of Newton polyhedrons in the group $(\mathbb{C}^*)^n$. This formula (which is multi-dimensional generalization of Vieta formula) uses Parshin symbols. Its proof however is based on simple geometry and combinatorics and does not use Parshin theory.

1.4. The logarithmic functional

The search for formula for the product of the roots of a system of equations convinced me that for $K = \mathbb{C}$ there should be intuitive geometric explanation of Parshin symbols and reciprocity laws. Such explanation based on multidimensional logarithmic functional was finally found in Ref. 6.

This paper is devoted to one-dimensional case. *One-dimensional logarithmic functional* (word "one-dimensional" we will omit further) associates each one-dimensional cycle γ in one-dimensional complex X and piecewise-smooth mapping $(f, g) : X \to (\mathbb{C}^*)^2$ into group $(\mathbb{C}^*)^2$ with a complex number, defined up to an integer additive term (i.e., with an element of the group $\mathbb{C}/\mathbb{Z}$). Logarithmic functional is direct generalization of usual logarithm. *Zero-dimensional logarithmic functional* (see Section 7.1) reduces to the usual logarithmic function. All theorems about logarithmic functional translate almost automatically to multidimensional case. I started presentation from one-dimensional case because of the following reasons:

1) The Weil reciprocity law is formulated much simpler than Parshin reciprocity laws. That's why the introduction of the logarithmic functional is much clearer in one-dimensional case.

2) Properties of logarithmic functional and methods of their prove become apparent enough already in one-dimensional case.

3) In one-dimensional case there already exist simple Beilinson functional giving clear topological proof of the reciprocity law. It has only one drawback: it is not transparent how to generalize it for multi-dimensional case. One of the goals of this publication is to compare logarithmic functional and Beilinson functional. The later is defined for a pair of rational functions f, g and one-dimensional cycle γ on complex algebraic curve X. Generalizing Beilinson functional we define logarithmic functional in Beilinson form, or in short LB-*functional*, for a pair of smooth functions f, g and one-dimensional cycle on real manifold M.

LB-functional has many properties of Beilinson functional, however there are some differences too: when functions f, g are fixed, LB-functional not always gives a class of one-dimensional cohomology group of the manifold M. For this it is required that the form $df \wedge dg$ is identical to zero on the manifold M. For the rational functions f, g on complex curve X the identity $df \wedge dg \equiv 0$ holds automatically. This fact plays a key role in the proof of the reciprocity law with the help of LB-functional.

We show that logarithmic functional is always representable as LB-functional. This representation relates logarithmic functional with Beilinson functional.

1.5. *Organization of material*

In Section 2 the Weil reciprocity law is given. In section 3 LB-functional is defined for a pair of complex-valued function on a circle. In Section 4 LB-functional is defined for a mapping $(f, g) : M \to (\mathbb{C}^*)^2$ of the manifold M into group $(\mathbb{C}^*)^2$ and closed oriented curve on manifold M. Section 5 discusses Beilinson's proof of the Weil theorem. In Section 6 LB-functional is defined for the mapping $(f, g) : M \to (\mathbb{C}^*)^2$ and one-cycle on the manifold M, being an image of one-cycle γ in one-dimensional complex X of piecewise-smooth mapping $\phi : X \to M$. Section 7 defines logarithmic functional and proves its main properties. Section 8 shows that logarithmic functional can always be represented in the form of LB-functional.

2. Formulation of the Weil reciprocity law

Let Γ be a connected compact one-dimensional complex manifold (another words, Γ is irreducible regular complex algebraic curve). *Local parameter* u near point $a \in \Gamma$ is defined as an arbitrary meromorphic function u with zero of multiplicity 1 at the point a. Local parameter is a coordinate function in small neighborhood of the point a.

Let φ be a meromorphic function on the curve Γ and $\sum_{k \leq m} c_m u^m$ be a Laurent series with respect to local parameter u near point a. We will call the *leading monomial* the first nonzero term of the series, i.e. $\chi(u) = c_k u^k$. The leading monomial is defined for any meromorphic function φ not identical to zero.

For every pair of meromorphic functions f, g not identical to zero on a curve Γ, and every point $a \in \Gamma$ the Weil symbol $[f, g]_a$ is defined. It is nonzero complex number given by

$$[f, g]_a = (-1)^{nm} a_m^n b_n^{-m},$$

where $a_m u^m$ and $b_n u^n$ are leading monomials on parameter u of functions f and g at the point a. Weil symbol is defined with the help of parameter u but *it does not depend on the choice of this parameter.*

Let v be another local parameter near point a and let cv be the leading monomial of the function u on parameter v, i.e. $u = cv + \ldots$. Then $a_m c^m v^m$ and $b_n c^n v^n$ are leading monomials of functions f and g on parameter v.

The equality $a_m^n b_n^{-m} = (a_m c^m)^n (b_n c^n)^{-m}$ proves the correctness of the definition of Weil symbol.

As it is seen from the definition *Weil symbol multiplicatively depends on functions f and g.* Multiplicativity on f means that if $f = f_1 f_2$ then $[f, g]_a = [f_1 g]_a [f_2 g]_a$. Multiplicativity with respect to g is defined analogously.

For every pair of meromorphic functions f, g not identical to zero on a curve Γ, the Weil symbol $[f, g]_a$ differs from one only in finite set of points a. Indeed *Weil symbol can differ from one only on the union of supports o the divisors of functions f and g.*

The Weil reciprocity law. (see ...) For every pair of meromorphic functions f, g not identical to zero on an irreducible algebraic curve Γ, the equality

$$\prod [f, g]_a = 1$$

holds. Here product is taken over all points a of the curve Γ.

Infinite product above makes sense since only finite number of terms in it are different from one.

Simple algebraic prove of the law can be found in Ref. 4. Simple topological proof based on properties of LB-functional described in Sections 3 and 4, can be found in Section 5 (this proof is reformulation of Beilinson's reasoning from Ref. 1).

3. LB-functional of the pair of complex valued functions of the segment on real variable

Let J be a segment $a \leq x \leq b$ of the real line. For any continuous on segment J function having non-zero complex values $f : J \to \mathbb{C}^*$, where $\mathbb{C}^* = \mathbb{C} \setminus 0$, denote by $\ln f$ any continuous branch of the logarithm of f. Function $\ln f$ is defined up to an additive term $2k\pi i$, where k is an integer. If $f(a) = f(b)$ *then we can define an integer number* $\deg_J \frac{f}{|f|}$ *– degree of mapping* $\frac{f}{|f|} : J/\partial J \to S^1$ *of the circle $J/\partial J$, obtained from segment J by identifying its ends a and b, to the unit circle $S^1 \subset \mathbb{C}$.* Obviously, $\deg_J \frac{f}{|f|} = \frac{1}{2\pi i}(\ln f(b) - \ln f(a))$.

Consider a pair of piecewise-smooth complex-valued functions f and g, having no zero values on the segment J, and $f(a) = f(b)$, $g(a) = g(b)$. For such pair of functions we will call LB-functional the complex number

$LB_J(f,g)$ given by formula

$$LB_J(f,g) = \frac{1}{(2\pi i)^2} \int_a^b \ln f \frac{dg}{g} - \frac{1}{2\pi i} \deg_J \frac{f}{|f|} \ln g(b).$$

Lemma 3.1. *LB-functional is defined up to an integer additive term and is well defined element of the group $\mathbb{C}/\mathbb{Z}$.*

Proof. Function $\ln f$ is defined up to an additive term $2k\pi i$, $\int_a^b \frac{dg}{g} = 2\pi i \deg_J \frac{dg}{|g|}$, and $\deg_J \frac{dg}{|g|}$ is an integer number. This means that the number $\frac{1}{(2\pi i)^2} \int_a^b \ln f \frac{dg}{g}$ is defined up to an integer additive term. The value $\ln g(b)$ is defined up to an additive term $2m\pi i$ and $\deg_J \frac{df}{|f|}$ is an integer number. This means that the number $-\frac{1}{2\pi i} \deg_J \frac{f}{|f|} \ln g(b)$ is defined up to an integer. $\square$

Now we give slightly more general formula for LB-functional. Let $a = x_0 < x_1 \cdots < x_n = b$ be an increasing sequence of points on the segment $[a,b]$. Given a continuous function $f : J/\partial J \to \mathbb{C}^*$ we construct a discontinuous function ϕ, which is equal to one of continuous branches $\ln_j f$ of the logarithm of f on each interval J_j defined by inequalities $x_j < x < x_{j+1}$. Let *jump* of function ϕ at point x_j be an integer number $m_\phi(x_j) = \frac{1}{2\pi i}(\lim_{t \to x_j^+} \phi(x) - \lim_{t \to x_j^-} \phi(x))$ for $j = 1, \ldots, n-1$, and at point x_n – an integer number $m_\phi(x_n) = \frac{1}{2\pi i}(\lim_{t \to x_n^+} \phi(x) - \lim_{t \to x_0^-} \phi(x))$.

Lemma 3.2. *Let for functions f, g and segment J LB-functional be defined. Then the equality*

$$LB_J(f,g) = \frac{1}{(2\pi i)^2} \int_J \phi \frac{dg}{g} - \frac{1}{2\pi i} \sum_{i=1}^n m_\phi(x_i) \ln g(x_i)$$

holds.

Proof. First show that $LB_J(f,g) =$

$$\frac{1}{(2\pi i)^2} \left(\sum_{i=0}^n \int_{J_i} \ln_i f \frac{dg}{g} + \sum_{i=0}^{n-1} \ln_i(x_i) \ln g(x_i) - \sum_{i=1}^n \ln_i(x_{i+1}) \ln g(x_i) \right).$$

Let's change function ϕ on interval J_j by adding to the branch of $\ln_j f$ number $2k\pi i$ leaving ϕ without changes on other intervals. As the result the value of LB-functional above is incremented by the number

$\frac{2k\pi i}{(2\pi i)^2}\left(\int_{J_i}\frac{dg}{g}+\ln g(x_i)-\ln g(x_{i+1})\right)$, which is equal to zero. Since the change of the branch of the logarithm on any of the intervals J_j does not affect the result, we can assume that the function ϕ is taken as a continuous branch of the logarithm of function f on the whole segment $J=[a,b]$. In this case the formula for the LB-functional holds (it coincides with the definition of the LB-functional). The claim of lemma follows, since $2\pi im_\phi(x_j)=\ln_j f(x_j)-\ln_{j+1} f(x_j)$ for $j=1,\ldots,n-1$, and $2\pi im_\phi(x_n)(\phi)=\ln_n f(x_n)-\ln_0 f(x_0)$ $\square$

If one of functions f or g is constant then it is easy to compute LB-functional. The following is obvious

Lemma 3.3. *Let for functions f, g and segment J LB-functional be defined. If $f\equiv C$ then $LB_J(f,g)=\frac{1}{2\pi i}\ln C\deg_J\frac{g}{|g|}$. If $g\equiv C$ then $LB_J(f,g)=-\frac{1}{2\pi i}\ln C\deg_J\frac{f}{|f|}$.*

The following obvious lemma shows that under change of variable LB-functional behaves as an integral of a differential form.

Lemma 3.4. *Let for functions f, g and segment J functional $LB_J(f,g)$ be defined, and let $\phi:J_1\to J$ be a piecewise-smooth homeomorphism of segment J_1 into segment J. Then functional $LB_{J_1}(f\circ\phi,g\circ\phi)$ is defined. Additionally, if ϕ preserves orientation then $LB_{J_1}(f\circ\phi,g\circ\phi)=LB_J(f,g)$, and if ϕ changes orientation then $LB_{J_1}(f\circ\phi,g\circ\phi)=-LB_J(f,g)$.*

Now we discuss how the LB-functional changes under homotopy of the pair of functions f,g. Let I be the unit segment $0\le t\le 1$ and $F:I\times J\to C^*$, $G:I\times J\to C^*$ be piecewise-smooth functions such, that $F(t,a)=F(t,b)$, $G(t,a)=G(t,b)$ for every fixed t. Let $f_0(x)=F(0,x)$, $f_1(x)=F(1,x)$, $g_0(x)=G(0,x)$, $g_1(x)=G(1,x)$.

Theorem 3.1. *The equality holds:*

$$LB_J(f_1,g_1)-LB_J(f_0,g_0)=\frac{1}{(2\pi i)^2}\int_{I\times J}\frac{dF}{F}\wedge\frac{dG}{G}.$$

Proof. Differential of the form $\ln F\frac{dG}{G}$ is equal to $\frac{dF}{F}\wedge\frac{dG}{G}$. Using Stokes formula we get

$$\int_{I\times J}\frac{dF}{F}\wedge\frac{dG}{G}=\int_a^b\ln f_1\frac{dg_1}{g_1}-\int_a^b\ln f_0\frac{dg_0}{g_0}-\int_0^1(\ln F(t,b)-\ln F(t,a))\frac{dG}{G}.$$

$$(1)$$

Since the degree of the mapping is homotopy invariant, the difference $\ln F(t, b) - \ln F(t, a)$ does not depend on parameter t and is equal to both numbers $2\pi i \deg_J \frac{f_1}{|f_1|}$ and $2\pi i \deg_J \frac{f_0}{|f_0|}$. Therefore

$$\int_0^1 (\ln F(t, b) - \ln F(t, a)) \frac{dG}{G} = 2\pi i \deg_J \frac{f_1}{|f_1|} \ln g_1(a) - 2\pi i \deg_J \frac{f_0}{|f_0|} \ln g_0(a)$$

(2)

and the statement of the theorem follows. □

Corollary 3.1. *Let f and g be piecewise-smooth functions on real line periodic with period $A = b - a$ with values in $\mathbb{C}^*$ and let J_c be a segment $a + c \leq x \leq b + c$ of the length A. Then $LB_{J_c}(f, g)$ does not depend on the choice of the point c.*

Proof. Consider homotopy $F(t, x) = f(t + x)$, $G(t, x) = g(t + x)$. This homotopy preserves LB-functional since $dF \wedge dG \equiv 0$. From previous theorem the proof follows. □

Lemma 3.5. *The equality $LB_J(f, g) = -LB_J(g, f)$ holds.*

Proof. Using equalities $\ln(b) = \ln(a) + 2\pi i \deg_J \frac{f}{|f|}$, $\ln g(b) = \ln g(a) + 2\pi i \deg_J \frac{g}{|g|}$ and Newton-Leibnitz formula we get:

$$\int_a^b d[\ln f \ln g] = 2\pi i \deg_J \frac{f}{|f|} \ln g(a) + 2\pi i \deg_J \frac{g}{|g|} \ln f(a) + \quad (3)$$

$$(2\pi i)^2 \deg_J \frac{f}{|f|} \deg_J \frac{g}{|g|}.$$

On the other hand

$$\int_a^b d[\ln f \ln g] = \int_a^b \ln g \frac{df}{f} + \int_a^b \ln f \frac{dg}{g}.$$

Two expressions for the integral $\int_a^b d[\ln f \ln g]$ must coincide. □

Lemma 3.6. *For any piecewise-smooth function $f : J \to \mathbb{C}^*$, having equal values at the end-points a and b of the segment J, $LB_J(f, f) = \frac{1}{2} \deg_J \frac{f}{|f|}$.*

Proof. Substituting $g = f$ in (3) we obtain

$$2 \int_a^b \frac{1}{2\pi i} \ln f \frac{df}{f} = 2 \deg_J \frac{f}{|f|} \ln f(c) + 2\pi i \deg_J^2 \frac{f}{|f|}.$$

The claim follows, since $\deg_J^2 \frac{f}{|f|} \equiv \deg_J \frac{f}{|f|} \mod 2$.   □

Lemma 3.7. *For any three piecewise-smooth functions f, φ, g with values in the group $\mathbb{C}^*$ on segment J with end-points a and b, such that $f(a) = f(b)$, $\varphi(a) = \varphi(b)$, $g(a) = g(b)$, the equality*

$$LB_J(f\varphi, g) = LB_J(f, g) + LB_J(\varphi, g)$$

holds.

Proof. In order to prove this it is enough to use the following facts:
1) in group $\mathbb{C}/\mathbb{Z}$ the equality $\frac{1}{2\pi i} \ln(f\varphi) = \frac{1}{2\pi i} \ln f + \frac{1}{2\pi i} \ln \varphi$ holds.
2) for any pair of continuous functions f and φ that do not have zero values and $f(a) = f(b)$, $\varphi(a) = \varphi(b)$, the equality $\deg_J \frac{f\varphi}{|f\varphi|} = \deg_J \frac{f}{|f|} + \deg_J \frac{\varphi}{|\varphi|}$ holds.   □

The following lemma is obvious and we only give a formulation of it.

Lemma 3.8. *Let segment J with end-points a and b is split by a point c $(a < c < b)$ into two segments: J_1 with end-points a, c and J_2 with end-points c, b. Let f, g be a pair of piecewise-smooth functions on J with values in group $\mathbb{C}^*$ such, that $f(a) = f(c) = f(b)$, $g(a) = g(c) = g(b)$. Then the functionals $LB_J(f, g)$, $LB_{J_1}(f, g)$, $LB_{J_2}(f, g)$ are defined and*

$$LB_J(f, g) = LB_{J_1}(f, g) + LB_{J_2}(f, g).$$

4. *LB*-functional of the pair of complex valued functions and one-dimensional cycle on real manifold

Let M be a smooth real manifold and $K(M)$ – multiplicative group with elements being smooth complex-valued functions on M not having values of 0. Let $\gamma : J \to M$ be piecewise-smooth closed curve on M. Element $I_J(\gamma^* f, \gamma^* g)$ of the group $\mathbb{C}/\mathbb{Z}$ will be called *LB-functional of the pair of functions $f, g \in K(M)$ and oriented closed curve γ* and will be denoted as $LB_\gamma(f, g)$.

Lemma 4.1. *LB-functional of the pair of functions $f, g \in K(M)$ and oriented closed curve γ does not change under orientation preserving reparametrization of the curve γ.*

Proof. For re-parametrization of the curve γ preserving the end-point $\gamma(a) = \gamma(b)$ the statement is proved in Lemma 3.4. Independence of the LB-functional from the choice of point $\gamma(a) = \gamma(b)$ is proved in Corollary 3.1. $\qquad\square$

An element $\mathrm{LB}_\gamma(f,g)$ of the group $\mathbb{C}/\mathbb{Z}$, defined by formula $\mathrm{LB}_\gamma(f,g) = \sum k_i \mathrm{LB}_{\gamma_i}(f,g)$, is called *LB-functional of pair of functions* $f,g \in K(M)$ *and cycle* $\gamma = \sum k_i \gamma_i$, where $k_i \in \mathbb{Z}$ and γ_i parameterized closed curves on manifold M.

For any cycle γ in manifold M and any function $f \in K(M)$ denote by $\deg_\gamma \frac{f}{|f|}$ the degree of mapping $\frac{f}{|f|} : \gamma \to S^1$ of the cycle γ into unit circle S^1.

Theorem 4.1. *For any cycle γ in the manifold M the following equalities hold:*

1) $LB_\gamma(f,g) = -LB_\gamma(g,f)$ for any pair of functions $f,g \in K(M)$.
2) $LB_\gamma(f\varphi,g) = LB_\gamma(f,g) + LB_\gamma(\varphi,g)$ for any $f,\varphi,g \in K(M)$.
3) $LB_\gamma(f,f) = \frac{1}{2}\deg_\gamma \frac{f}{|f|}$ for any $f \in K(M)$.
4) $LB_\gamma(C,f) = -LB_\gamma(f,C) = \ln C \deg_\gamma \frac{f}{|f|}$ for any function $f \in K(M)$ and any non-zero constant C.

Proof. Statement 1) follows from Lemma 3.5, statement 2) follows from Lemma 3.7, statement 3) – from Lemma 3.6 and statement 4) – from Lemma 3.3. $\qquad\square$

Theorem 4.2. *If for a pair of functions $f,g \in K(M)$ on manifold M the equality $df \wedge dg \equiv 0$ holds, then LB-functional $LB_\gamma(f,g)$ depends only on homology class of the cycle $\gamma = \sum k_i \gamma_i$.*

Proof. According to Theorem 3.1 under conditions of Theorem 4.2 LB-functional along closed curve does not change under homotopy of the curve. By homotopying if necessary every component of the cycle γ it is possible to assume that cycle consists from closed curves, passing through the fixed point c. Consider the fundamental group $\pi_1(M,c)$ of the manifold M with basis point c. For functions f and g satisfying condition of the theorem, the mapping $I : \pi_1(M,c) \to \mathbb{C}/\mathbb{Z}$ of the closed curve $\gamma \in \pi_1(M,c)$ into element $I_\gamma(f,g)$ of group $\mathbb{C}/\mathbb{Z}$ is defined correctly and is group homomorphism. Any homomorphism of the fundamental group into Abel group is passed trough the homomorphism of the fundamental group into group of one-dimensional homologies. $\qquad\square$

5. Topological proof of the Weil reciprocity law

Beilinson's functional is a particular case of LB-functional, in which f, g is a pair of analytic functions on one-dimensional complex manifold, and γ is one-dimensional real cycle on this manifold. Under this restrictions we will call LB-functional Beilinson's functional and write $B_\gamma(f, g)$ instead of $LB_\gamma(f, g)$.

Example 5.1. (Beilinson, see Ref. 1) Let M be one-dimensional complex manifold and $K_a(M) \subset K(M)$ be a subgroup of group $K(M)$ consisting of analytic functions not equal to 0 anywhere. For any two functions $f, g \in K_a(M)$ the equality $df \wedge dg \equiv 0$ holds. In accordance with Theorem 4.2, for any pair of functions $f, g \in K_a(M)$ one-dimensional co-chain on manifold M, that associates with a cycle γ element $B_\gamma(f, g) \in \mathbb{C}/\mathbb{Z}$, is an element of $H^1(M_a, \mathbb{C}/\mathbb{Z})$. According to Theorem 4.1, this class has the following properties:

1) $B_\gamma(f, g) = -B_\gamma(g, f)$ for any pair of functions $f, g \in K_a(M_a)$.
2) $B_\gamma(f\varphi, g) = B_\gamma(f, g) + B_\gamma(\varphi, g)$ for any $f, \varphi, g \in K_a(M_a)$.
3) $B_\gamma(f, f) = \frac{1}{2} \deg_\gamma \frac{f}{|f|}$ for any $f \in K_a(M_a)$.
4) $B_\gamma(C, f) = -B_\gamma(f, C) = \ln C \deg_\gamma \frac{f}{|f|}$ for any function $f \in K_a(M_a)$ and arbitrary nonzero constant C.

Let X be a connected compact one-dimensional complex manifold with a boundary, and f, g be nonzero meromorphic functions on the manifold X, that are regular on the boundary $\gamma = \partial X$ of manifold X and not equal to zero in the points of boundary ∂X. Under these assumptions the following lemma holds.

Lemma 5.1. *Let a single valued branch of the function $\ln f$ exists on the manifold X. Then B-functional $B_\gamma(f, g)$ of functions f, g and of boundary of the manifold X is equal to $\sum\limits_{p \in X} \operatorname{ord}_a g \frac{1}{2\pi i} \ln f(a)$, where $\operatorname{ord}_a g$ is order of meromorphic function g at the point a. (The infinite sum appearing above is well defined since only finitely many terms in this sum are not equal to zero.)*

Proof. The statement follows from the equality

$$\frac{1}{(2\pi i)^2} \int\limits_{\partial X} \ln f \frac{dg}{g} = \frac{1}{2\pi i} \sum_{p \in X} (\operatorname{ord}_p g) \ln f(p). \qquad \square$$

Let $U \subset \mathbb{C}$ be a simply-connected domain with smooth boundary γ, containing point $0 \in U$. Let f, g be meromorphic functions in the domain U, such that their restrictions on punctured domain $U \setminus 0$ are analytic functions, not taking values of 0. Let $a_1 z^k, a_2 z^m$ be leading terms of Laurent series of functions f and g at the point 0: $f = a_1 z^k + \ldots$, and $g = a_2 z^m + \ldots$.

Lemma 5.2. *Under the listed above assumptions*

$$B_\gamma(f, g) = \frac{km}{2} + m \ln a_1 - k \ln a_2.$$

Proof. Represent f, g as $f = f_1 z^k$, $g = g_1 z^m$, where f_1, g_1 are analytic in the domain U. Observe, that $f_1(0) = a_1 \neq 0$, $g_1(0) = a_2 \neq 0$. Using multiplicativity property of B-functional we get $B_\gamma(f, g) = km B_\gamma(z, z) + m B_\gamma(f_1, z) + k B_\gamma(z, g_1) + B_\gamma(f_1, g_1)$. Using Lemma 5.1 and skew-symmetric property of B-functional we get $B_\gamma(f_1, z) = \ln a_1$, $B_\gamma(z, g_1) = -\ln a_2$ and $B_\gamma(f_1, g_1) = 0$. Further $B_\gamma(z, z) = \frac{1}{2} \deg_\gamma \frac{z}{|z|} = \frac{1}{2}$. $\square$

Topological proof of the Weil theorem is actually completed. We just need to reformulate obtained results. Group $\mathbb{C}/\mathbb{Z}$ is isomorphic to the multiplicative group $\mathbb{C}^*$ of the field of the complex numbers. Required isomorphism is given by mapping $\tau : \mathbb{C}/\mathbb{Z} \to \mathbb{C}^*$ defined by formula $\tau(a) = \exp(2\pi i a)$. For any one-dimensional cycle γ in the manifold M and for any pair of functions $f, g \in K(M)$ we call *exponential B-functional* $\tilde{B}_\gamma(f, g)$ an element of the group $\mathbb{C}^*$, defined by formula $\tilde{B}_\gamma(f, g) = \exp(2\pi i B_\gamma(f, g))$.

Using the notion of exponential B-functional the last Lemma can be reformulated as follows

Lemma 5.3. *Under assumptions of Lemma 5.2, exponential B-functional* $\tilde{B}_\gamma(f, g)$ *coincides with the Weil symbol* $[f, g]_0$ *of functions f and g at point 0.*

Let Γ be a compact complex curve, f, g – meromorphic functions on Γ such that they are not identical to zero on any connected component of Γ, D – a union of supports of divisors of functions f and g. Let U_p be a small open disk containing point $p \in D$, and $\gamma_p = \partial U_p$ be the border of disk U_p, $U = \bigcup_{p \in D} U_p$, $\gamma = \sum_{p \in D} \gamma_p$. Consider manifold $W = M \setminus U$. Cycle γ homologically equals to zero on manifold W, since $\partial W = -\gamma$. Therefore, the equality $B_\gamma(f, g) = 0$ holds and $\exp(2\pi i B_\gamma(f, g)) = \prod_{p \in P} \exp(2\pi i B_{\gamma_p}(f, g)) = 1$. By Lemma 5.3 we have $\exp(2\pi i B_{\gamma_p}(f, g)) = [f, g]_p$ which proves the Weil theorem.

6. Generalized LB-functional

In this section we will give another (more general, but in fact equivalent) definition of LB-functional. It will be easier to generalize to multidimensional case. Any one-dimensional cycle in $(\mathbb{C}^*)^2$ (as in any other manifold) can be viewed as integer combination of oriented closed curves, i.e. as integer combination of images of oriented circle. Therefore, it is sufficient to define LB-functional for a pair of functions on a standard circle, as it was done above. When $n > 1$ n-dimensional cycle in manifold M can be viewed as an image of the mapping $F : X \to M$ of some n-dimensional cycle γ in some n-dimensional simplicial complex X. Here we give a definition of LB-functional for an image in the group $(\mathbb{C}^*)^2$ of one-dimensional cycle, laying in one-dimensional simplicial complex X under mapping $F : X \to (\mathbb{C}^*)^2$.

Let X be one-dimensional simplicial complex. Let us fix orientation on each edge of complex X. Denote by $S(\Delta_j, Q_p)$ the incidence coefficient between edge Δ_j and vertex Q_p. It is equal to zero if and only if vertex Q_p does not belong to the edge Δ_j, and it is equal to $+1$ or -1 depending on the sign of Q_p as a boundary of oriented edge Δ_j.

For any continuous function $f : X \to \mathbb{C}^*$ on edges of complex X it is possible to choose single-valued branches of function $\ln f$. We will need a definition of jump function for such a collection of branches of function $\ln f$ on one-dimensional cycle $\gamma = \sum k_j \Delta_j$ of complex X. Let a single-valued branch $\ln_j f : \Delta_j \to \mathbb{C}^*$ of the multi-valued function $\ln f$ be fixed on the edge Δ_j of complex X. Denote by ϕ a collection of branches $\ln f_j$. We will view collection ϕ as discontinuous function on complex X (the collection ϕ defines a function on the complement of complex X to the set of its vertices, but at vertices of complex X this function is multi-valued). Define on vertices of complex X jump function for function ϕ and cycle γ. If vertex Q_p is adjacent to the edge Δ_j then on the vertex Q_p a restriction of function $\ln_j f$ given on the edge Δ_j is defined. We define the value of the jump function $m_{\phi,\gamma}$ for function ϕ and cycle $\gamma = \sum k_j \Delta_j$ at the vertex Q_p of complex X by formula

$$m_{\phi,\gamma}(Q_p) = \frac{1}{2\pi i} \sum S(\Delta_j, Q_p) k_j \ln_j f(Q_p),$$

where summation is done over all edges Δ_j of complex X.

Lemma 6.1. *The jump function is an integer valued function on the vertices of complex X.*

Proof. On every edge Δ_j adjacent to the vertex Q_p we can choose branches $\ln_j f$ so they are equal at Q_p. With such choice of the function ϕ the number

$m_{\phi,\gamma}(Q_p)$ is equal to zero, because the chain $\gamma = \sum k_j \Delta^j$ is a cycle. With the change of branches $\ln_j f$ the change in the value of jump function is integer number. □

Consider piecewise-smooth mapping $(f, g) : X \to (\mathbb{C}^*)^2$ of one-dimensional complex X and one-dimensional cycle $\gamma = \sum k_j \Delta_j$ in X. Let $\phi = \{\ln_j f\}$ be a collection of single-valued branches of function $\ln f$ on edges Δ_j of complex X. For pair of functions f, g and cycle γ we will call LB-functional the complex number $\mathrm{LB}_\gamma(f, g)$ defined by formula

$$\mathrm{LB}_\gamma(f, g) = \frac{1}{(2\pi i)^2} \int_\gamma \phi \frac{dg}{g} - \frac{1}{2\pi i} \sum_{Q \in V} m_{\phi,\gamma}(Q) \ln g(Q),$$

where summation is over set V of all vertices of complex X and $m_{\phi,\gamma}$ is jump function for function ϕ and cycle γ.

The following lemmas a proved similarly to Lemma 3.1 and 3.2.

Lemma 6.2. *LB-functional $\mathrm{LB}_\gamma(f, g)$ is defined up to an integer additive term and is correctly defined element of the group $\mathbb{C}/\mathbb{Z}$.*

Lemma 6.3. *LB-functional $\mathrm{LB}_\gamma(f, g)$ for the mapping $(f, g) : X \to (\mathbb{C}^*)^2$ does not change under a simplicial subdivision of the complex X.*

Let discuss the change of LB-functional under homotopy of the mapping (f, g). Let I be a unit segment $0 \le t \le 1$ and $(F, G) : I \times X \to (\mathbb{C}^*)^2$ be a piecewise-smooth mapping, which coincides with the mapping $(f_0, g_0) : X \to (\mathbb{C}^*)^2$ when $t = 0$ and with the mapping $(f_1, g_1) : X \to (\mathbb{C}^*)^2$ when $t = 1$. Similarly to the Theorem 3.1 we can prove

Theorem 6.1. *The equality*

$$\mathrm{LB}_\gamma(f_1, g_1) - \mathrm{LB}_\gamma(f_0, g_0) = \frac{1}{(2\pi i)^2} \int_{I \times \gamma} \frac{dF}{F} \wedge \frac{dG}{G}$$

holds.

7. Logarithmic function and logarithmic functional

In this section we define logarithmic functional, formulate and prove its major properties. Definitions and results of this section can be easily extended to the multi-dimensional case and we plan to return to this topic in future publications.

7.1. Zero-dimensional logarithmic functional and logarithm

In order to highlight an analogy between logarithm and logarithmic functional we will define zero-dimensional functional here.

Consider group $\mathbb{C}^*$ with coordinate z. Group $\mathbb{C}^*$ is homotopy-equivalent to the circle T^1, defined by the equation $|z| = 1$. Denote by $e = 1$ the unit element of the group $\mathbb{C}^*$.

Let X be finite set of points and $K(X)$ – multiplicative group with elements being complex-valued functions on X having no 0 values. Thus element f of the set $K(X)$ is a mapping $f : X \to \mathbb{C}^*$.

Let D be zero-dimensional cycle from the group $H_0(X, \mathbb{Z})$, i.e. D is a linear combination of points from the set X with integer coefficients: $D = \sum k_j x_j$, where $k_j \in \mathbb{Z}$ i $x_j \in X$.

Any mapping $f : X \to (\mathbb{C}^*)^2$ of the finite set X into group $\mathbb{C}^*$ is homotopic to the mapping of X into point e since group $\mathbb{C}^*$ is connected.

We will call *zero-dimensional logarithmic functional* a functor that associates with the pair f, D, consisting of function $f \in K(M)$ and zero-dimensional cycle $D = \sum k_j x_j$, the complex number

$$\ln(f, D) = \frac{1}{2\pi i} \int\limits_\gamma \frac{dz}{z},$$

where γ is arbitrary piecewise-smooth curve in group $\mathbb{C}^*$ with the boundary $\partial\gamma$ equal to $\sum k_j f(x_j) - (\sum k_j)e$.

Obviously, the number $\ln(f, D)$ is defined up to an additive integer term and is correctly defined element of the group $\mathbb{C}/\mathbb{Z}$. The following formula holds

$$\ln(f, D) = \frac{1}{2\pi i} \sum k_j \ln f(x_j),$$

where $\ln f(x_j)$ is any of values of multi-valued function $\ln f$ at point x_j.

Thus zero-dimensional logarithmic functional reduces to the usual logarithmic function. It has the following multiplicativity property: for any pair of functions $f, g \in K(X)$ and any zero-dimensional cycle D the equality $\ln(fg, D) = \ln(f, D) + \ln(g, D)$ holds. For any fixed function $f \in K(X)$ and any pair of zero-dimensional cycles D_1, D_2 the equality $\ln(f, (D_1 + D_2)) = \ln(f, D_1) + \ln(f, D_2)$ holds. Another words, for fixed function f functional $\ln(f, D)$ is zero-dimensional co-chain with the values in group $\mathbb{C}/\mathbb{Z}$.

Let M be real manifold and $f : X \to \mathbb{C}^*$ be a continuous mapping. Let's associate with every zero-dimensional cycle $D = \sum k_j x_j$ ($k_j \in \mathbb{Z}$, $x_j \in M$) on the manifold M an element $\ln(f, D)$ of the group $\mathbb{C}/\mathbb{Z}$ defined by formula $\ln(f, D) = \frac{1}{2\pi i} \sum k_j \ln f(x_j)$. It is obvious that *co-chain* $\ln(f, D)$ *defines an element of zero-dimensional cohomology group of manifold M with coefficients in group $\mathbb{C}/\mathbb{Z}$ if and only if $df \equiv 0$* (i.e. if and only if the function f is constant on each connected component of the manifold M).

7.2. *Properties of one-dimensional logarithmic functional*

In this subsection we define one-dimensional logarithmic functional and formulate its important properties. As before we will use the term "logarithmic functional" skipping word "one-dimensional".

Consider group $(\mathbb{C}^*)^2$ with co-ordinate functions z_1 and z_2. Group $(\mathbb{C}^*)^2$ is homotopy equivalent to the torus $T^2 \subset (\mathbb{C}^*)^2$, defined by equations $|z_1| = |z_2| = 1$. On group $(\mathbb{C}^*)^2$ there is remarkable 2-form

$$\omega = \frac{1}{(2\pi i)^2} \frac{dz_1}{z_1} \wedge \frac{dz_2}{z_2}.$$

Restriction of form ω to the torus T^2 is real 2-form

$$\omega|_{T^2} = \frac{1}{4\pi^2} d(\arg z_1) \wedge d(\arg z_2),$$

that is not equal to zero nowhere. Integral of form ω over the torus T^2, oriented by the form $d(\arg z_1) \wedge d(\arg z_2)$, is equal to 1. Define by Id a subset of group $(\mathbb{C}^*)^2$ consisting of points (z_1, z_2) such, that one of co-ordinates is equal to 1, i.e. $Id = \{z_1 = 1\} \cup \{z_2 = 1\}$.

Let X be one-dimensional simplicial complex and let $\gamma = \sum k_j \Delta_j$ be integer linear combination of its oriented edges Δ_j which constitutes a cycle, i.e. $\partial \gamma = 0$. Let $(f, g) : X \to (\mathbb{C}^*)^2$ be a piecewise-smooth mapping of complex X into the group $(\mathbb{C}^*)^2$.

Logarithmic functional is a functor, that associates with the mapping $(f, g) :$ $X \to (\mathbb{C}^*)^2$ and one-dimensional cycle γ on X an element $\ln(f, g, \gamma)$ of the group $\mathbb{C}/\mathbb{Z}$, defined by formula

$$\ln(f, g, \gamma) = \frac{1}{(2\pi i)^2} \int_\sigma \frac{dz_1}{z_1} \wedge \frac{dz_2}{z_2} = \int_\sigma \omega,$$

where σ is 2-chain in $(\mathbb{C}^*)^2$, with the boundary $\partial \sigma$ equal to the difference of the image $\mathbf{f}_*(\gamma)$ of cycle γ under mapping $\mathbf{f} = (f, g)$ and some cycle γ_1 lying in the set Id.

We list the major properties of the logarithmic functional:

1) The value of the logarithmic functional is correctly defined element of the group $\mathbb{C}/\mathbb{Z}$.

2) Logarithmic functional skew-symmetrically depends on components f

and g, i.e. $\ln(f,g,\gamma) = -\ln(g,f,\gamma)$.

3) Logarithmic functional multiplicatively depends on components f and g, i.e.

$$\ln(f\phi,g,\gamma) = \ln(f,g,\gamma) + \ln(\phi,g,\gamma), \ln(f,g\phi,\gamma) = \ln(f,g,\gamma) + \ln(f,\phi,\gamma).$$

4) Logarithmic functional has the following topological property. Let M be a real manifold and $(f,g) : M \to \mathbb{C}^*$ be a smooth mapping. One-dimensional cycle $\tilde{\gamma}$ on manifold M can be viewed as an image under some piecewise-smooth mapping $\phi : X \to M$ of some one-dimensional cycle $\gamma = \sum k_j \Delta_j$ on some one-dimensional complex X, where $k_j \in \mathbb{Z}$ and Δ_j are edges of complex X. Associate with the cycle $\tilde{\gamma}$ on manifold M an element $\ln(f \circ \phi, g \circ \phi, \gamma)$ of group $\mathbb{C}/\mathbb{Z}$. Obtained with this association one-dimensional co-chain gives a class of one-dimensional cohomology group of manifold M with coefficients in group $\mathbb{C}/\mathbb{Z}$ if and only if $df \wedge dg \equiv 0$ (i.e. if and only if the differential of the mapping $(f,g) : M \to (\mathbb{C}^*)^2$ degenerates at each point of manifold M).

5) The equality $\ln(f,g,\gamma) = \mathrm{LB}_\gamma(f,g)$ holds. One one hand, this equality gives a formula for logarithmic functional that does not use auxiliary 2-chain σ. On the other hand, this equality gives geometric sense to LB-functional and shows that LB-functional is analog of logarithm.

7.3. *Prove of properties of logarithmic functional*

This subsection will give prove of properties 1)–4). Property 5) will be proved in next section.

It is easy to see that sets Id (defined in previous subsection) and $(\mathbb{C}^*)^2$ have the following topological properties:

1) The set Id is homotopy equivalent to the boucquet of two circles, $\pi_1(Id)$ – free group with two generators and $H_1(Id, \mathbb{Z}) \approx \mathbb{Z} + \mathbb{Z}$.

2) The set $(\mathbb{C}^*)^2$ is homotopy equivalent to the real torus T^2 and $\pi_1((\mathbb{C}^*)^2) \approx H_1((\mathbb{C}^*)^2, \mathbb{Z}) \approx \mathbb{Z} + \mathbb{Z}$.

3) Enclosure $\pi : Id \to (\mathbb{C}^*)^2$ induces isomorphism of one-dimensional ho-mologies and mapping "onto" of the fundamental groups of sets Id and $(\mathbb{C}^*)^2$.

It follows from item 3) that *any continuous mapping* $f : X \to (\mathbb{C}^*)^2$ *of one-dimensional simplicial complex* X *into group* $(\mathbb{C}^*)^2$ *is homotopic to the mapping of complex* X *into the set* $Id \subset (\mathbb{C}^*)^2$. We now prove more precise statement.

Let V be a set of vertices of one-dimensional complex X and E be the set of its edges. Fix inside each edge $\tau_j \in E$ two different points A_j and B_j. Points A_j, B_j split edge τ_j into three segments, which intersect at the ends only — *central* segment with endpoints A_j and B_j and two *boundary* segments. Fix also in each edge τ_j point O_j, laying inside central segment $[A_j, B_j]$.

Denote by X_1 the union of all central segments on all edges τ_j, and by X_0 – the union of all extreme segments on all edges τ_j. As it is seen from this construction, every connected component of the set X_1 is a central segment of one of the edges τ_j. Every connected component of the set X_0 contains one of the vertices of complex X and is the union of all extreme segments, containing this vertex. So, every connected component of the sets X_0 and X_1 is contractible.

Lemma 7.1. *Any continuous (piecewise-smooth) mapping $(f, g) : X \to (\mathbb{C}^*)^2$ of one-dimensional complex X into group $(\mathbb{C}^*)^2$ is homotopic (piecewise-smoothly homotopic) to the mapping $(f_1, g_1) : X \to (\mathbb{C}^*)^2$ such, that function f_1 is equal to 1 on the set X_1 and function g_1 is equal to 1 on the set X_0.*

Proof. First we will homotopically change function f without changing function g. Since the set $\mathbb{C}^*$ is connected we can assume that the following condition is satisfied: $f(O_j) = 1$ at the chosen point O_j on every edge τ_j. If this is not so, then we can homotopically change function f in such a way, that in the process of homotopy f_t values $f_t(O_j)$ move along the curves connecting points O_j with 1. Let this condition holds. It is possible to construct a homotopy $\phi_t : X \to X$ of the identity mapping of the complex X into itself such, that it leaves intact every vertex of the complex, translates every edge τ_j into itself and contracts every central segment $[A_j, B_j]$ into the point O_j. Homotopy $f_t = f \circ \phi_t$ translates function f into function f_1 which is equal to 1 on the set X_1. Analogously, without changing function f, we can homotopically change function g into the function g_1 having required properties. We can assume, that function g is equal to 1 at every vertex (otherwise it can be homotopically changed into the function having this property). It is possible to construct homotopy $\psi_t : X \to X$ of the identity mapping of the complex X into itself such, that it leaves intact every vertex of the complex, translates every edge τ_j into itself and contracts each of boundary segments into the vertex that is contained in this segment. Homotopy $g_t = g \circ \psi_t$ translates function g into function g_1 which is equal to 1 on the set X_0. $\square$

On group $(\mathbb{C}^*)^2$ there is remarkable 2-form

$$\omega = \frac{1}{(2\pi i)^2} \frac{dz_1}{z_1} \wedge \frac{dz_2}{z_2}.$$

Lemma 7.2. *For any 2-cycle σ in $(\mathbb{C}^*)^2$ integral $\int_\sigma \omega$ is an integer number.*

Proof. Consider projection $\rho : (\mathbb{C}^*)^2 \to T^2$ of the group $(\mathbb{C}^*)^2$ into torus T^2, given by formula $\rho(z_1, z_2) = (\frac{z_1}{|z_1|}, \frac{z_2}{|z_2|})$. It is clear, that integral $\int_\sigma \omega$ is equal to the degree of the mapping $\rho : \sigma \to T^2$, which is restriction of projection ρ onto cycle σ. Thus, this integral is an integer number.   □

Theorem 7.1. *The value $\ln(f, g, \gamma)$ of the logarithmic functional is correctly defined element of the group $\mathbb{C}/\mathbb{Z}$ (i.e. when 2-chain σ in the definition of functional changes, the value can be incremented by integer number only).*

Proof. Since embedding $\pi : Id \to (\mathbb{C}^*)^2$ induces isomorphism π_* of groups $H_1(Id, \mathbb{Z})$ and $H_1((\mathbb{C}^*)^2, \mathbb{Z})$, for the image $(f, g)_* \gamma$ of cycle γ there exists homological cycle γ_1 laying in the set Id, i.e. there exists 2-chain σ_1 such, that $(f, g)_* \gamma - \gamma_1 = \partial \sigma_1$. Let γ_2 be another cycle in Id that is homological to the cycle $(f, g)_* \gamma$ and σ_2 be such 2-chain, that $(f, g)_* \gamma - \gamma_2 = \partial \sigma_2$. Since isomorphism π_* has no kernel, cycles γ_1 and γ_2 are homological in the set Id, i.e. there exists a chain σ_3 laying in Id such, that $\gamma_1 - \gamma_2 = \partial \sigma_3$. By the construction the chain $\sigma = \sigma_2 - \sigma_1 - \sigma_3$ has zero boundary and therefore is a 2-cycle in $(\mathbb{C}^*)^2$. By Lemma 7.2 integral $\int_\sigma \omega$ is an integer number. On the other hand $\int_\sigma \omega = \int_{\sigma_2} \omega - \int_{\sigma_1} \omega - \int_{\sigma_3} \omega$. The chain σ_3 is laying in the set Id. Restriction of the form ω on Id is identically equal to zero, therefore $\int_{\sigma_3} \omega = 0$. Thus, the difference $\int_{\sigma_2} \omega - \int_{\sigma_1} \omega$ is an integer number.   □

Theorem 7.2. *Logarithmic functional is skew-symmetrical, i.e. $\ln(f, g, \gamma) = -\ln(g, f, \gamma)$.*

Proof. Let $R : (\mathbb{C}^*)^2 \to (\mathbb{C}^*)^2$ be a mapping that swaps co-ordinates, i.e. let $R(z_1, z_2) = (z_2, z_1)$. Under this mapping the form ω changes the sign, i.e. $R^* \omega = -\omega$. Mapping R translates the set Id into itself. Let σ be a 2-chain in $(\mathbb{C}^*)^2$ such, that the cycle $(\partial \sigma) - (f, g)_*(\gamma)$ is contained in the set Id. Then $\ln(f, g, \gamma) = \int_\sigma \omega$. Further, $\ln(g, f, \gamma) = \int_{R(\sigma)} \omega = \int_\sigma R^* \omega = -\int_\sigma \omega = -\ln(f, g, \gamma)$.   □

Theorem 7.3. *Logarithmic functional has the following multiplicative properties:*

$$\ln(f\varphi, g, \gamma) = \ln(f, g, \gamma) + \ln(\varphi, g, \gamma), \quad \ln(f, g\varphi, \gamma) = \ln(f, g, \gamma) + \ln(f, \varphi, \gamma).$$

Proof. Since the functional is skew-symmetrical, it is sufficient to prove the first equality $\ln(f\varphi, g, \gamma) = \ln(f, g, \gamma) + \ln(\varphi, g, \gamma)$. For evaluation of $\ln(f, g, \gamma)$ and $\ln(\varphi, g, \gamma)$ we will choose specific 2-chains σ_1 and σ_2. Let $X = X_0 \cup X_1$ be a covering of the complex X by two closed sets with contractible connected components the same as in Lemma 7.1.

We describe the choice of the chain σ_1 first. Let $I = [0, 1]$ be a unit segment, $W = I \times X$ and $(F, G) : W \to (\mathbb{C}^*)^2$ be a piecewise-smooth homotopy of the mapping (f, g) discussed in Lemma 7.1. Another words, let (F, G) be such mapping, that: 1) restriction $(F, G)|_{\{1\} \times X}$ of the mapping (F, G) on the set $\{1\} \times X$ coincides with the mapping (f, g) when sets X and $\{1\} \times X$ are identified, 2) restriction $F|_{\{0\} \times X_1}$ of function F on the set $\{0\} \times X_1$ identically equals to 1, 3) restriction $G|_{\{0\} \times X_0}$ of function G on the set $\{0\} \times X_0$ identically equals to 1. For the cycle $\gamma = \sum k_i \Delta_i$ take $\sigma_1 = \sum k_i (F, G)_* (I \times \Delta_i)$. By construction the boundary of the chain σ_1 is equal to $(f, g)_*(\gamma) - \gamma_1$, where γ_1 is a cycle laying in the set Id. Thus $\ln(f, g, \gamma) = \int_{\sigma_1} \omega$.

Chain σ_2 is constructed analogously. Let $(\Phi, G) : W \to (\mathbb{C}^*)^2$ be piecewise-smooth mapping, with component G the same as in above described homotopy, and component Φ having the following properties: 1) restriction $\Phi|\{1\} \times X$ coincides with function φ, 2) restriction $\Phi|_{\{0\} \times X_1}$ of function Φ on the set $\{0\} \times X_1$ identically equals to 1. For cycle $\gamma = \sum k_i \Delta_i$ take $\sigma_2 = \sum k_i (\Phi, G)_* (I \times \Delta_i)$. By construction the boundary of the chain σ_2 is equal to $(\varphi, g)_*(\gamma) - \gamma_2$, where γ_2 is a cycle laying in the set Id. Thus $\ln(\varphi, g, \gamma) = \int_{\sigma_2} \omega$.

Now we construct chain σ_3. Consider the mapping $(F\Phi, G) : W \to (\mathbb{C}^*)^2$ with first component being the product of functions F and Φ, and second component equal to function G. Take $\sigma_3 = \sum k_i (F\Phi, G)_* (I \times \Delta_i)$. By construction the boundary of the chain σ_3 is equal to the difference of the cycle $(f\varphi, g)_*(\gamma)$ and the cycle laying in the set Id. Therefore, $\ln(f\varphi, g, \gamma) = \int_{\sigma_3} \omega$. Thus we have the following equalities

$$\ln(f, g, \gamma) = \frac{1}{2\pi i} \sum k_i \int_{I \times \Delta_i} \frac{dF}{F} \wedge \frac{dG}{G},$$

$$\ln(\varphi, g, \gamma) = \frac{1}{2\pi i} \sum k_i \int\limits_{I \times \Delta_i} \frac{d\Phi}{\Phi} \wedge \frac{dG}{G},$$

$$\ln(f\varphi, g, \gamma) = \frac{1}{2\pi i} \sum k_i \int\limits_{I \times \Delta_i} \frac{d(F\Phi)}{(F\Phi)} \wedge \frac{dG}{G}.$$

Equality $\ln(f\varphi, g, \gamma) = \ln(f, g, \gamma) + \ln(\varphi, g, \gamma)$ now follows from the identity

$$\frac{d(F\Phi)}{(F\Phi)} = \frac{dF}{F} + \frac{d\Phi}{\Phi}. \qquad \qquad \square$$

Let real manifold M and smooth mapping $(f, g) : M \to (\mathbb{C}^*)^2$ be given. With each triple, consisting of one-dimensional complex X, one-dimensional cycle γ on X and piecewise-smooth mapping $\phi : X \to M$, we will associate an element $\ln(f \circ \phi, g \circ \phi, \gamma)$ of group $\mathbb{C}/\mathbb{Z}$. In order to prove topological property 4) of the logarithmic functional described in previous subsection it is sufficient to prove the following theorem.

Theorem 7.4. *Let form $df \wedge dg$ be identically equal to zero on the manifold M and let cycle $\phi_*\gamma$ homologically equals to zero. Then $\ln(f \circ \phi, g \circ \phi, \gamma) = 0$. Conversely, if from homological equality to zero of the cycle $\phi_*\gamma$ follows equality $\ln(f \circ \phi, g \circ \phi, \gamma) = 0$ then on the manifold M form $df \wedge dg$ is identically equal to zero.*

Proof. If cycle $\phi_*\gamma$ homologically equals to zero, then on the manifold M there exists piecewise-smooth 2-chain σ such that $\partial\sigma = \phi \circ \gamma$. By definition

$$\ln(f \circ \phi, g \circ \phi, \gamma) = \frac{1}{(2\pi i)^2} \int\limits_{\sigma} \frac{df}{f} \wedge \frac{dg}{g}.$$

If form $df \wedge dg$ is identically equal to zero, then integrand is equal to zero. Thus, $\ln(f \circ \phi, g \circ \phi, \gamma) = 0$.

Conversely, let on some bi-vector v at some point a of manifold M form $df \wedge dg$ is not equal to zero. Consider arbitrary smooth mapping $\phi : (\mathbb{R}^2, 0) \to (M, a)$ of the standard plane $\mathbb{R}^2$ into manifold M that translates point 0 into point $a \in M$, and differential $d\phi$ is nonsingular at point 0 and maps plane $\mathbb{R}^2$ into subspace of the tangent space TM_a at the point a, containing bi-vector v. Let S_ε be a square $-\varepsilon \le x \le \varepsilon, -\varepsilon \le y \le \varepsilon$ on the plane $\mathbb{R}^2$, and $X_\varepsilon = \partial S_\varepsilon$ be its boundary. Let γ be a cycle geometrically coinciding with X_ε and oriented as boundary of the square S_ε. Obviously, for small enough ε element $\ln(f \circ \phi, g \circ \phi, \gamma)$ of group $\mathbb{C}/\mathbb{Z}$ is not equal to zero, however cycle $\phi_*\gamma$ by construction is homologically equal to zero. $\qquad \square$

8. Logarithmic functional and generalized LB-functional

In this section we prove the following theorem.

Theorem 8.1. *The equality* $\ln(f, g, \gamma) = LB_\gamma(f, g)$ *holds.*

We first prove auxiliary result.

Lemma 8.1. *Let $X_0, X_1 \subset X$ be subsets of complex X defined in Lemma 7.1, $(f, g) : X \rightarrow (\mathbb{C}^*)^2$ – piecewise-smooth mapping such, that restriction of function f on X_1 and restriction of function g on X_0 identically equal to 1. Then for any one-dimensional cycle γ in complex X the equality $LB_\gamma(f, g) = 0$ holds.*

Proof. Consider the following subpartition of complex X: every edge τ_j of complex X is split into three edges by points A_j, B_j belonging to the boundaries of sets X_0 and X_1. Choose collection ϕ of the branches of function $\ln f$ on the edges of sub-partitioned complex X that satisfies the following two conditions: 1) on every edge that is connected component of set X_1, the branch of function $\ln f$ is identically equal to zero; 2) branches $\ln f$ on the union of edges, that are connected components of the set X_0, define on this component continuous function. Condition 2) can be satisfied because every connected component of the set X_0 is contractible. By the definition of LB-functional we have

$$LB_\gamma(f, g) = \frac{1}{(2\pi i)^2} \int_\gamma \phi \frac{dg}{g} - \frac{1}{2\pi i} \sum_{P \in \tilde{V}} m_{\phi,\gamma}(P) \ln g(P),$$

where summation is over set $\tilde{V}$ of all vertices of sub-partitioned complex X. Now, $\int_\gamma \phi \frac{dg}{g} = 0$, because integrand is equal to zero. Indeed, on every edge from set X_1 function ϕ is equal to zero. On every edge from set X_0 function g is identically equal to 1 and thus $dg/g = 0$. At the vertex P of the original complex X (before subpartition), coefficient $m_{\phi,\gamma}$ is equal to 0, because function ϕ is continuous at such point P. At every vertex P, belonging to the boundary of sets X_0 and X_1, the number $\ln g$ is equal to $2k\pi$, where $k \in \mathbb{Z}$, because function g is equal to 1 on the set X_0. Therefore, the number $\frac{1}{2\pi i} \sum_{P \in \tilde{V}} m_{\phi,\gamma}(P) \ln g(P)$ is integer. From all this it follows that element $LB_\gamma(f, g)$ of group $\mathbb{C}/\mathbb{Z}$ equals to zero. $\square$

Proof. (of Theorem 8.1). By Lemma 7.1 the mapping $(f, g) : X \rightarrow (\mathbb{C}^*)^2$ is homotopic to the mapping $(f_1, g_1) : X \rightarrow (\mathbb{C}^*)^2$, for which restriction of function f_1 on X_1 and restriction of function g_1 on X_0 are identically equal to 1. Let $W = I \times X$ and $(F, G) : W \to (\mathbb{C}^*)^2$ be homotopy connecting mappings (f, g) and (f_1, g_1). By Theorem 6.1

$$\mathrm{LB}_\gamma(f, g) - \mathrm{LB}_\gamma(f_1, g_1) = \int_\sigma \omega,$$

where σ — 2-chain on W that is equal to $I \times \gamma$. By Lemma 8.1 $\mathrm{LB}_\gamma(f_1, g_1) = 0$. By construction $\partial\sigma = \gamma - \gamma_1$, where cycle γ_1 is laying in the set Id. Therefore, $\mathrm{LB}_\gamma(f, g) = \ln_\gamma(f, g)$.   □

The author would like to thank Eugene Zima for his help during preparation of this paper.

References

1. Beilinson A.A. Higher regulators and values of L-functions of curves. *Functional Analysis and its Applications*, 14(2):116–118, 1980.
2. Deligne P. Le symbole modéré. Publications Mathématiques de l'IHÉS, 73 (1991), p. 147-181
3. Fimmel T., Parshin A. N. Introduction to higher adelic theory. Preprint, Moscow, 1999.
4. Khovanskii A. Newton polytopes, curves on toric surfaces, and inversion of weil's theorem. *Russian Math. Surveys*, 52(6):1251–1279, 1997.
5. Khovanskii A. Newton polyhedrons, a new formula for mixed volume, product of roots of a system of equations. In *Proceed. of a Conf. in Honour of V.I. Arnold, Fields Inst. Comm. vol 24*, Amer. Math. Soc., pages 325–364, USA, 1999.
6. Khovanskii A. A Function *Log* in the Space of n-Dimensional Cycles and Parshin–Kato Theory. *International Conference on Toric Topology*, Osaka City University, Osaka, Japan, May-June 2006.
   http://higeom.math.msu.su/people/taras/TTC/slides/Khovanskii/
7. McLaughlin D.A. Brylinski J.-L. Multidimensional reciprocity laws. *J. reine angew. Math.*, 481:125–147, 1996.
8. Parshin A. N. *Local class field theory*, Trudy Mat. Inst. Steklov, volume 165. 1984.
9. Parshin A. N. *Galois cohomology and Brauer group of local fields*, Trudy Mat. Inst. Steklov, volume 183. 1984.

ON SOLUTIONS OF LINEAR FUNCTIONAL SYSTEMS AND FACTORIZATION OF LAURENT–ORE MODULES

MIN WU

Software Engineering Institute, East China Normal University
3663 North Zhangshan Road, Shanghai 200062, China
and
Key Lab of Mathematics–Mechanization, Chinese Academy of Sciences
Zhong Guan Cun, Beijing 100080, China
E-mail: mwu@sei.ecnu.edu.cn

ZIMING LI

Key Lab of Mathematics–Mechanization, Chinese Academy of Sciences
Zhong Guan Cun, Beijing 100080, China
E-mail: zmli@mmrc.iss.ac.cn

We summarize some recent results on partial linear functional systems. By associating a finite-dimensional linear functional system to a Laurent–Ore module, Picard–Vessiot extensions are generalized from linear ordinary differential (difference) equations to finite-dimensional linear functional systems. A generalized Beke's method is also presented for factoring Laurent–Ore modules and it will allow us to find all "subsystems" whose solution spaces are contained in that of a given linear functional system.

Keywords: Linear functional systems; Fully integrable systems; Laurent–Ore algebras; Modules of formal solutions; Picard–Vessiot extensions; Fundamental matrices; Eigenrings.

1. Introduction

This paper provides a survey of the work by M. Bronstein and the authors in a *France–Sino Scientific Cooperation Project*[a] from 2002 to 2005. Our project concerns finite-dimensional linear functional systems, and its outcome includes: a generalization of Picard–Vessiot extensions of linear

---

[a]This joint project is supported in part by the French Government Scholarship (BGF no. 2002915), an INRIA-CAFÉ Project Funding and two National Key Projects of China (no. 1998030600, 2004CB31830).

ordinary differential (difference) equations, the notion of modules of formal
solutions, algorithms for computing the dimension of solution spaces and
for reducing linear functional systems to fully integrable ones, and generalizations of Beke's factorization algorithm and of the eigenring method. The
emphasis of this paper is on descriptions of these results. Precise references
are given for proofs and technical details.

A (partial) linear functional system consists of linear partial differential,
shift, and q-shift operators, or any mixture thereof. By a finite-dimensional
linear functional system, or a ∂-finite system for short, we mean a linear
functional system whose module of formal solutions has finite dimension
(see Definition 4.1). Intuitively, a system is ∂-finite if and only if its solution
space has finite dimension. Below is an example:

$$
\begin{cases}
P''(x, k) - \frac{2x}{1-x^2} P'(x, k) + \frac{k(k+1)}{1-x^2} P(x, k) & = 0 \\
P(x, k+2) - \frac{(2k+3)x}{k+2} P(x, k+1) + \frac{k+1}{k+2} P(x, k) = 0 \, .
\end{cases}
\tag{1}
$$

The sequence of the Legendre polynomials $\{P(x, k)\}_{k=1}^{\infty}$ is a solution of (1)
with the initial conditions

$$\{ P(0,0) = 0, \quad P'(0,0) = 0, \quad P(0,1) = 0, \quad P'(0,1) = 1 \}.$$

Given a linear functional system L, we are interested in the following questions: (i) Does L have a nonzero solution? (ii) Is there a ring containing
"all" the solutions of L? (iii) How does one compute the dimension of the
solution space of L? (iv) How does one find (if it exists) a "subsystem"
whose solution space is properly contained in that of L? (v) Determine
whether the solution space of L can be written as a direct sum of those of
its subsystems?

Our work is intended for answering these questions algorithmically for ∂-
finite systems. In terms of modules of formal solutions (Definition 4.1) and
Picard–Vessiot extensions (Definition 4.2), the above questions translate
respectively to: (i) Is a module M of formal solutions trivial? (ii) Does there
exist a Picard–Vessiot extension for a given system? (see Section 4) (iii) How
does one compute the dimension of M? (see Section 5) (iv) How does one
find a nontrivial submodule of M? (see Section 6) (v) Is M decomposable?
(see Section 6)

Many of the results in this paper are straightforward generalizations of their
counterparts from linear ordinary differential or difference equations. These
generalizations are however necessary in view of their wider applicability
and the complications caused by the appearance of several differential and
difference operators.

Throughout the paper, rings are not necessarily commutative and have arbitrary characteristic. Ideals, modules and vector spaces are all left ones. Fields are always assumed to be commutative. Denote by $R^{p \times q}$ the set of all $p \times q$ matrices with entries in a ring R, and by $\mathbf{e}_{in}$, for $1 \leq i \leq n$, the unit vector in $R^{1 \times n}$ with 1 in the ith position and 0 elsewhere. The notation "$\cong_R$" means "isomorphic as R-modules". We use $(\cdot)^\tau$ to denote the transpose of a vector or matrix, and $\mathbf{1}_n$ to denote the identity matrix of size n. Vectors are represented by the boldfaced letters $\mathbf{u}, \mathbf{v}, \mathbf{w}$ etc. Vectors of unknowns are denoted $\mathbf{x}, \mathbf{y}, \mathbf{z}$, etc. The symbol $\mathbb{C}$ denotes the field of complex numbers.

The paper is organized as follows. In Section 2, we present some preliminaries and define the notion of linear functional systems. In Section 3, we construct Picard–Vessiot extensions for fully integrable systems, which are a common special case of ∂-finite systems. In Section 4, modules of formal solutions are defined and Picard–Vessiot extensions are generalized for ∂-finite systems. In Section 5, we present some techniques for computing linear dimension of a linear functional system. In Section 6, we generalize Beke's algorithm and the eigenring approach to factor Laurent–Ore modules. Concluding remarks are made in Section 7.

2. Preliminaries

Let R be a ring and Δ be a finite set of commuting maps from R to itself. A map in Δ is assumed to be either a derivation or an automorphism. Recall that a derivation δ is an additive map satisfying the multiplicative rule $\delta(ab) = a\delta(b) + \delta(a)b$ for all $a, b \in R$. The pair (R, Δ) is called a Δ-*ring*, and it is a Δ-*field* when R is a field.

For a derivation $\delta \in \Delta$, an element c of R is called a *constant* with respect to δ if $\delta(c) = 0$. For an automorphism $\sigma \in \Delta$, c is called a *constant* with respect to σ if $\sigma(c) = c$. An element c of R is called a *constant* if it is a constant with respect to all maps in Δ. The set of constants of R, denoted by C_R, is a subring. The ring C_R is a subfield if R is a field.

Let (F, Δ) be a Δ-field. By reordering the indices, we can always assume that $\Delta = \{\delta_1, \ldots, \delta_\ell, \sigma_{\ell+1}, \ldots, \sigma_m\}$ for some $\ell \geq 0$, where the δ_i's are derivation operators on F and the σ_j's are automorphisms of F. The *Ore algebra*[8] over F is the polynomial ring $\mathcal{S} := F[\partial_1, \ldots, \partial_m]$ in ∂_i with the usual addition and the multiplication defined as follows:

$$\partial_i \partial_j = \partial_j \partial_i, \quad \partial_s a = a \partial_s + \delta_s(a), \quad \partial_t a = \sigma_t(a) \partial_t,$$

for any $1 \leq i, j \leq m$, $1 \leq s \leq \ell$, $\ell < t \leq m$ and $a \in F$.

Remark that $\partial_i(a)$, where a is an element of a Δ-ring, is meant to be $\delta_i(a)$ if ∂_i is associated to a derivation operator δ_i, and to be $\sigma_i(a)$ if ∂_i is associated to an automorphism σ_i; while $\partial_i a$, where a is an element of the Ore algebra $\mathcal{S}$, means the product of ∂_i and a.

Definition 2.1. Let (F, Δ) be a Δ-field. A *linear functional system* over F is a system of the form $A(\mathbf{z}) = 0$ where A is a $p \times q$ matrix with entries in the Ore algebra $\mathcal{S}$ and $\mathbf{z}$ is a column vector of q unknowns.

Example 2.1. The system (1), satisfied by the Legendre polynomials, can be rewritten as $A(z) = 0$ where

$$A = \left(\partial_x^2 - \frac{2x}{1 - x^2} \partial_x + \frac{k(k + 1)}{1 - x^2}, \; \partial_k^2 - \frac{(2k + 3)x}{k + 2} \partial_k + \frac{k + 1}{k + 2} \right)^\tau,$$

with ∂_x the differentiation with respect to x and ∂_k the shift operator with respect to k.

Let F be a Δ-field. A commutative ring R containing F is called a Δ-*extension* of F if all the maps in Δ can be extended to R in such a way that all derivations (resp. automorphisms) of F become derivations (resp. automorphisms) of R and the extended maps commute pairwise.

By a solution of a linear functional system $A(\mathbf{z}) = 0$ over F, we mean a vector $(s_1, \ldots, s_q)^\tau$ over some Δ-extension of F such that $A(s_1, \ldots, s_q)^\tau = 0$, *i.e.*, the application of the matrix A to the vector is zero.

3. Fully integrable systems

A common special case of linear functional systems consists of fully integrable systems, which are of the form $\{\partial_i(\mathbf{z}) = A_i \mathbf{z}\}_{1 \le i \le m}$ and correspond to the linear functional system $A(\mathbf{z}) = 0$ where the matrix A is given by the stacking of blocks of the form $(\partial_i \cdot \mathbf{1}_n - A_i)$. Fully integrable systems are of interest to our study, since to every ∂-finite system, we can associate a fully integrable system whose solution space is isomorphic to that of the original system. (see Section 4.3)

Definition 3.1. A system of the form

$$\delta_i(\mathbf{z}) = A_i \mathbf{z}, \quad 1 \le i \le \ell, \qquad \sigma_i(\mathbf{z}) = A_i \mathbf{z}, \quad \ell + 1 \le i \le m, \qquad (2)$$

where $A_i \in F^{n \times n}$ and $\mathbf{z}$ is a column vector of n unknowns, is called an *integrable system* of size n over F if the following compatibility conditions are satisfied:

$$\delta_i(A_j) = \delta_j(A_i), \qquad\qquad 1 \le i < j \le \ell,$$
$$\sigma_i(A_j)A_i = \sigma_j(A_i)A_j, \qquad \ell < i < j \le m, \qquad (3)$$
$$\sigma_j(A_i)A_j = A_iA_j + \delta_i(A_j), \quad 1 \le i \le \ell < j \le m.$$

The integrable system (2) is said to be *fully integrable* if the matrices $A_{\ell+1}, \ldots, A_m$ are invertible.

Using Ore algebra notation, we write $\{\partial_i(\mathbf{z}) = A_i\mathbf{z}\}_{1 \le i \le m}$ for the system (2) where the action of ∂_i is again meant to be δ_i for $i \le \ell$ and to be σ_i for $i > \ell$. Observe that the conditions (3) are derived from the condition $\partial_i(\partial_j(\mathbf{z})) = \partial_j(\partial_i(\mathbf{z}))$ and are exactly the matrix-analogues of the compatibility conditions for first-order scalar equations in Ref. 10. For a linear ordinary difference equation, we often assume that its trailing coefficient is nonzero, while, for a first-order matrix difference equation, we assume that its matrix is invertible. These assumptions lead to the condition on invertibility of $A_{\ell+1}, \ldots, A_m$ in Definition 3.1.

Example 3.1. Let $F = \mathbb{C}(x, k)$, δ_x be the differentiation with respect to x and σ_k the shift operator with respect to k. Then

$$\mathcal{A} : \{\, \delta_x(\mathbf{z}) = A_x\mathbf{z}, \ \sigma_k(\mathbf{z}) = A_k\mathbf{z} \,\}$$

is a fully integrable system where

$$A_x = \begin{pmatrix} \frac{x^2-kx-k}{x(x-k)(x-1)} & \frac{x^2-kx+3k-2x}{kx(x-k)(x-1)} \\ \frac{k(kx+x-x^2-2k)}{(x-k)(x-1)} & \frac{x^3+x^2-kx^2-2x+2k}{x(x-k)(x-1)} \end{pmatrix}$$

and

$$A_k = \begin{pmatrix} \frac{k+1+kx^2-xk^2-x}{(x-k)(x-1)} & -\frac{k+1+kx-k^2-x}{k(x-k)(x-1)} \\ \frac{x(k+1)(k+1+kx-k^2-x)}{(x-k)(x-1)} & \frac{(k+1)(x^2-2kx-x+k^2)}{k(x-k)(x-1)} \end{pmatrix}.$$

In what follows, we generalize fundamental matrices and Picard–Vessiot extensions of linear ordinary differential (difference) equations to fully integrable systems.

A square matrix with entries in a commutative ring is said to be *invertible* if its determinant is a unit in that ring.

Let F be a Δ-field and $\{\partial_i(\mathbf{z}) = A_i\mathbf{z}\}_{1 \le i \le m}$ be a fully integrable system of size n over F. We define

Definition 3.2. An $n \times n$ matrix U with entries in a Δ-extension of F is a *fundamental matrix* for the system $\{\partial_i(\mathbf{z}) = A_i\mathbf{z}\}_{1 \le i \le m}$ if U is invertible and $\partial_i(U) = A_i U$ for each i, *i.e.*, each column of U is a solution of the system.

A two-sided ideal I of a commutative Δ-ring R is said to be *invariant* if $\delta_i(I) \subset I$ for $i \leq \ell$ and $\sigma_j(I) \subset I$ for $j > \ell$. The ring R is said to be *simple* if its only invariant ideals are (0) and R.

Definition 3.3. A *Picard–Vessiot ring* for a fully integrable system is a (commutative) ring E such that:

(i) E is a simple Δ-extension of F.
(ii) There exists some fundamental matrix U with entries in E for the system such that E is generated by the entries of U and $\det(U)^{-1}$ over F.

Definitions 3.2 and 3.3 are natural generalizations of their analogues in the purely differential case [22, (pages 12, 415)] and the ordinary difference case [21, (Errata)].

The existence of fundamental matrices and Picard–Vessiot extensions for fully integrable systems is stated in the following

Theorem 3.1. *Every fully integrable system over F has a Picard–Vessiot ring E. If F has characteristic 0 and C_F is algebraically closed, then $C_E = C_F$. Furthermore, that extension is minimal, meaning that no proper subring of E satisfies both conditions in Definition 3.3.*

A detailed proof of the above theorem is found in Ref. 6.

Consequently, if F has characteristic zero and an algebraically closed field of constants, then all the solutions of a fully integrable system in its Picard–Vessiot ring form a C_F-vector space whose dimension equals the size of the system.

We now present two examples for Picard–Vessiot extensions. The reader is referred to [24, Section 2.2] for detailed verifications.

Example 3.2. Consider the fully integrable system of size one:

$$\partial_i(z) = a_i z \quad \text{where } a_i \in F \text{ and } i = 1, \dots, m.$$

This is an extension of Example 1.19 in Ref. 22.

Let $E = F[T, T^{-1}]$ be the Δ-extension such that $\delta_i(T) = a_i T$ for $i \leq \ell$ and $\sigma_j(T) = a_j T$ for $j > \ell$. Then E is a Picard–Vessiot ring of the given system if there does not exist an integer $k > 0$ and a nonzero $r \in F$ such that $\delta_i(r) = k a_i r$ for $i \leq \ell$ and $\sigma_j(r) = a_j^k r$ for $j > \ell$. Otherwise, assume that the integer $k > 0$ is minimal so that there exists a nonzero $r \in F$ satisfying $\delta_i(r) = k a_i r$ for $i \leq \ell$ and $\sigma_j(r) = a_j^k r$ for $j > \ell$. Then $E/(T^k - r)$ is a Picard–Vessiot ring of the given system.

Example 3.3. Consider the system $\mathcal{A}$ in Example 3.1. Note that the change of variable[b] $\mathbf{z} = M\mathbf{y}$ where

$$M = \begin{pmatrix} \frac{x-k}{x} & x^2 \\ (x-k)k & x^2 k \end{pmatrix},$$

transforms $\mathcal{A}$ into another fully integrable system

$$\mathcal{B} : \{\, \delta_x(\mathbf{y}) = B_x \mathbf{y}, \ \sigma_k(\mathbf{y}) = B_k \mathbf{y} \,\}$$

with

$$B_x = \begin{pmatrix} 1 & 0 \\ 0 & 0 \end{pmatrix} \quad \text{and} \quad B_k = \begin{pmatrix} 1 & 0 \\ 0 & k \end{pmatrix}.$$

It suffices to find a Picard–Vessiot ring of $\mathcal{B}$. We get that

$$V = \begin{pmatrix} e^x & 0 \\ 0 & \Gamma(k) \end{pmatrix}$$

is a fundamental matrix for $\mathcal{B}$, and thus MV is for $\mathcal{A}$. Moreover, $F[e^x, \Gamma(k), e^{-x}, \Gamma(k)^{-1}]$ is a Picard–Vessiot extension for $\mathcal{A}$.

4. ∂-finite systems

In this section, we first discuss generic solutions of linear algebraic equations over arbitrary rings, then introduce the notions of Laurent–Ore algebras and modules of formal solutions. These two notions allow us to generalize the results in Section 3 to ∂-finite systems.

4.1. *Generic solutions of linear algebraic equations*

Let R be an arbitrary ring. Denote by $Z(R)$ the *center* of R, *i.e.* the set of all elements that commute with every element in R. Then $Z(R)$ is a subring of R. Consider a $p \times q$ matrix $A = (a_{ij})$ with entries in R. For any R-module N, we can associate to A a $Z(R)$-linear map $\lambda : N^q \to N^p$ given by

$$\xi := (\xi_1, \ldots, \xi_q)^\tau \mapsto A\xi = \left(\sum_{j=1}^q a_{1j}\xi_j, \ldots, \sum_{j=1}^q a_{pj}\xi_j \right)^\tau.$$

---

[b]which can be found, for example, by computing the hyperexponential solutions of the system[10,14,24]

We therefore say that $\xi \in N^q$ is a *solution* "*in N*" of the system $A(\mathbf{z}) = 0$ if $\lambda(\xi) = 0$, and write $\mathrm{sol}_N(A(\mathbf{z}) = 0)$ for the set of all solutions in N. Clearly, $\mathrm{sol}_N(A(\mathbf{z}) = 0)$ is a $Z(R)$-module. Note that λ is in general not R-linear since R is noncommutative.

As in the case of $\mathcal{D}$-modules,[15] we can associate to $A \in R^{p \times q}$ an R-module as follows: A induces the R-linear map $\rho : R^{1 \times p} \to R^{1 \times q}$ given by $(r_1, \ldots, r_p) \mapsto (r_1, \ldots, r_p)A$. Let $M = R^{1 \times q}/(R^{1 \times p}A)$, which is the quotient of $R^{1 \times q}$ by the image of the map ρ. We call M the R-*cokernel* of A and denote it by $\mathrm{coker}_R(A)$. Clearly, $\mathrm{coker}_R(A)$ is an R-module. Let $\mathbf{e}_{1p}, \ldots, \mathbf{e}_{pp}$ and $\mathbf{e}_{1q}, \ldots, \mathbf{e}_{qq}$ be the canonical bases of $R^{1 \times p}$ and $R^{1 \times q}$, respectively. Denote by π the canonical map from $R^{1 \times q}$ to $\mathrm{coker}_R(A)$, and set $\mathbf{e}_j = \pi(\mathbf{e}_{jq})$ for $1 \le j \le q$. Since π is surjective, M is generated by $\mathbf{e}_1, \ldots, \mathbf{e}_q$ over R. Note that $\rho(\mathbf{e}_{ip})$ is the i-th row of A. Hence

$$0 = \pi(\rho(\mathbf{e}_{ip})) = \pi\left(\sum_{j=1}^q a_{ij}\mathbf{e}_{jq}\right) = \sum_{j=1}^q a_{ij}\pi(\mathbf{e}_{jq}) = \sum_{j=1}^q a_{ij}\mathbf{e}_j, \text{ for } 1 \le i \le p,$$

which implies that $(\mathbf{e}_1, \ldots, \mathbf{e}_q)^\tau$ is a solution of $A(\mathbf{z}) = 0$ in M.

Given two R-modules N_1 and N_2, denote by $\mathrm{Hom}_R(N_1, N_2)$ the set of all R-linear maps from N_1 to N_2. Clearly, $\mathrm{Hom}_R(N_1, N_2)$ is a $Z(R)$-module.

As illustrated by the following theorem, Proposition 1.1 of Ref. 15 remains true when $\mathcal{D}$ is replaced by an arbitrary ring R.

Theorem 4.1. *Let* $M = R^{1 \times q}/(R^{1 \times p}A)$. *Then for any R-module N,* $\mathrm{Hom}_R(M, N)$ *and* $\mathrm{sol}_N(A(\mathbf{z}) = 0)$ *are isomorphic as $Z(R)$-modules.*

Remark 4.1.

(i) The proof of Proposition 1.1 in Ref. 15 can be adapted to this theorem in a straightforward way (see Ref. 6) and also, an elementary proof is given in [24, Theorem 2.4.1].

(ii) The proof of Theorem 4.1 reveals that the vector $\mathbf{e} := (\mathbf{e}_1, \ldots, \mathbf{e}_q)^\tau \in M^q$ specified above is a "generic" solution of the system $A(\mathbf{z}) = 0$ in the sense that any solution $(s_1, \ldots, s_q)^\tau$ of that system in N is the image of $\mathbf{e}$ under the map in $\mathrm{Hom}_R(M, N)$ sending $\mathbf{e}_i$ to s_i.

4.2. *Laurent–Ore algebras*

Let F be a Δ-field and $\mathcal{S} = F[\partial_1, \ldots, \partial_m]$ be the corresponding Ore algebra. In the differential case, an $\mathcal{S}$-module is classically associated to a linear functional system.[15,22] In the difference case, however, $\mathcal{S}$-modules may not have appropriate dimensions, as illustrated by the following counterexample.

Example 4.1. Let $\sigma \neq 1$ be an automorphism of F and $\mathcal{S} = F[\partial]$ be the corresponding Ore algebra. The equation $\partial(y) = 0$ cannot have a fundamental matrix (u) in any difference ring extension of F, for otherwise, $0 = \partial(u) = \sigma(u)$, thus $u = 0$. Therefore $\partial(y) = 0$ has only trivial solution. However, the $\mathcal{S}$-module $\mathcal{S}/\mathcal{S}\partial$ has dimension one as an F-vector space.

In [21, page 56], modules over Laurent algebras are used instead to avoid the above problem. It is therefore natural to introduce the following extension of $\mathcal{S}$: let $\theta_{\ell+1}, \ldots, \theta_m$ be indeterminates independent of the ∂_i. Since the σ_j^{-1} are automorphisms of F, $\overline{\mathcal{S}} = F[\partial_1, \ldots, \partial_m, \theta_{\ell+1}, \ldots, \theta_m]$ is also an Ore algebra in which the θ_j are associated to the σ_j^{-1}. Note that $\partial_j \theta_j$ is in the center of $\overline{\mathcal{S}}$ since

$$(\partial_j \theta_j)a = \partial_j \sigma_j^{-1}(a)\theta_j = \sigma_j(\sigma_j^{-1}(a))\partial_j \theta_j = a\partial_j \theta_j,$$

for all $a \in F$ and $j > \ell$. Therefore the left ideal $T = \sum_{j=\ell+1}^m \overline{\mathcal{S}}(\partial_j \theta_j - 1)$ is a two-sided ideal of $\overline{\mathcal{S}}$, and we call the factor ring $\mathcal{L} = \overline{\mathcal{S}}/T$ the *Laurent–Ore algebra* over F. Identifying $\overline{\partial}_i$ with the image of ∂_i in $\mathcal{L}$ and writing ∂_j^{-1} for the image of θ_j in $\mathcal{L}$, we can write $\mathcal{L}$ (by convention) as $\mathcal{L} = F[\partial_1, \ldots, \partial_m, \partial_{\ell+1}^{-1}, \ldots, \partial_m^{-1}]$ and view it as an extension of $\mathcal{S}$. For linear ordinary difference equations, $\mathcal{L} = F[\sigma, \sigma^{-1}]$ is the algebra used in Ref. 21, while for linear partial difference equations with constant coefficients, $\mathcal{L}$ is the Laurent polynomial ring in Refs. 17,25.

Except for the purely differential case where $\ell = 0$, a Laurent–Ore algebra $\mathcal{L} - \Gamma[\partial_1, \ldots, \partial_m, \partial_{\ell+1}^{-1}, \ldots, \partial_m^{-1}]$ is not an Ore algebra since $\partial_j \partial_j^{-1} = \partial_j^{-1} \partial_j = 1$.

When revisiting Example 4.1 with Laurent–Ore algebras, we get that the left ideal generated by ∂ in $\mathcal{L} = F[\partial, \partial^{-1}]$ is $\mathcal{L}$, therefore the dimension of $\mathcal{L}/(\mathcal{L}\partial)$ over F, which is zero, equals that of the solution space of $\partial(y) = 0$ in any difference ring extension.

In the sequel, a module over a Laurent–Ore algebra that is finite-dimensional over the ground field is called a *Laurent–Ore module* for short.

4.3. *Modules of formal solutions*

Let F be a Δ-field, and $\mathcal{S}$ and $\mathcal{L}$ be the corresponding Ore and Laurent–Ore algebras. Replacing R with $\mathcal{L}$ in Theorem 4.1 yields

Theorem 4.2. *Let* $A \in \mathcal{S}^{p \times q}$ *and* $M=\mathrm{coker}_{\mathcal{L}}(A)$. *Then* $\mathrm{sol}_N(A(\mathbf{z})=0)$ *and* $\mathrm{Hom}_{\mathcal{L}}(M, N)$ *are isomorphic as* C_F-*vector spaces for any* $\mathcal{L}$-*module* N.

From Remark 4.1(ii) in which we replace arbitrary ring R with $\mathcal{L}$, $\mathrm{coker}_{\mathcal{L}}(A)$ describes the properties of all the solutions of $A(\mathbf{z}) = 0$ "anywhere". This motivates us to define

Definition 4.1. *Let* $A \in \mathcal{S}^{p \times q}$. *The* $\mathcal{L}$-*module* $M = \mathcal{L}^{1 \times q}/(\mathcal{L}^{1 \times p} A)$ *is called the* module of formal solutions *of the system* $A(\mathbf{z}) = 0$. *The dimension of* M *as an* F-*vector space is called the* linear dimension *of the system. The system is said to be* of finite linear dimension, *or simply,* ∂-finite, *if* $0 < \dim_F M < +\infty$.

Note that we choose to exclude systems with $\dim_F M = 0$ in the above definition since such system has only trivial solution in any $\mathcal{L}$-module, particularly, in any Δ-extension of F.

Remark 4.2. For any $A \in \mathcal{S}^{p \times q}$, we can construct both its $\mathcal{S}$-cokernel $\mathrm{coker}_{\mathcal{S}}(A)$ and $\mathcal{L}$-cokernel $\mathrm{coker}_{\mathcal{L}}(A)$. Viewing $\mathcal{L}$ as a right $\mathcal{S}$-module and $\mathrm{coker}_{\mathcal{S}}(A)$ as a left $\mathcal{S}$-module, we can define the tensor product[18] $\mathcal{L} \otimes_{\mathcal{S}} \mathrm{coker}_{\mathcal{S}}(A)$, which is a right $\mathcal{S}$-module and a left $\mathcal{L}$-module. Lemma 2.4.10 in Ref. 24 shows that $\mathrm{coker}_{\mathcal{L}}(A)$ and $\mathcal{L} \otimes_{\mathcal{S}} \mathrm{coker}_{\mathcal{S}}(A)$ are isomorphic as $\mathcal{L}$-modules. Thus $\dim_F \mathrm{coker}_{\mathcal{L}}(A)$ does not exceed $\dim_F \mathrm{coker}_{\mathcal{S}}(A)$.

Let $A(\mathbf{z}) = 0$ with $A \in \mathcal{S}^{p \times q}$ be a system of linear dimension n and M be its module of formal solutions with an F-basis $\mathbf{b}_1, \ldots, \mathbf{b}_n$. Suppose that $\partial_i(\mathbf{b}_1, \ldots, \mathbf{b}_n)^\tau = B_i(\mathbf{b}_1, \ldots, \mathbf{b}_n)^\tau$ where $B_i \in F^{n \times n}$ for $1 \leq i \leq m$. By a straightforward verification, $\{\partial_i(\mathbf{x}) = B_i\mathbf{x}\}_{1 \leq i \leq m}$ is a fully integrable system, which is called the *integrable connection* of $A(\mathbf{z}) = 0$ with respect to the basis $\mathbf{b}_1, \ldots, \mathbf{b}_n$ of M.

∂-finite and fully integrable systems are connected by the next proposition whose proof is given in [6, Proposition 2] and [24, Proposition 2.4.12].

Proposition 4.1. *Let* $A, \mathbf{b}_1, \ldots, \mathbf{b}_n, B_1, \ldots, B_m$ *be as above, and* B *be the stacking of the blocks* $(\partial_i \cdot \mathbf{1}_n - B_i)$. *Then*

(i) $\mathrm{coker}_{\mathcal{L}}(A) \cong_{\mathcal{L}} \mathrm{coker}_{\mathcal{L}}(B)$.

(ii) *Let* $\{\mathbf{e}_1, \ldots, \mathbf{e}_q\}$ *be the set of* $\mathcal{L}$-*generators of* M *satisfying* $A(\mathbf{e}_1, \ldots, \mathbf{e}_q)^\tau = 0$ *and* $P \in F^{q \times n}$ *be given by*

$$(\mathbf{e}_1, \ldots, \mathbf{e}_q)^\tau = P(\mathbf{b}_1, \ldots, \mathbf{b}_n)^\tau.$$

> *Then for any Δ-extension E of F, the correspondence $\xi \mapsto P\xi$ is an isomorphism of C_E-modules between $\mathrm{sol}_E(\{\partial_i(\mathbf{x}) = B_i\mathbf{x}\}_{1 \le i \le m})$ and $\mathrm{sol}_E(A(\mathbf{z}) = 0)$.*

Remark that the inverse of the correspondence in Proposition 4.1 (ii) is given by $\eta \mapsto Q\eta$, where Q is a matrix in $\mathcal{L}^{n \times q}$ such that

$$(\mathbf{b}_1, \ldots, \mathbf{b}_n)^\tau = Q(\mathbf{e}_1, \ldots, \mathbf{e}_q)^\tau.$$

From Proposition 4.1 (ii), all the solutions of the system $A(\mathbf{z}) = 0$ can be obtained from those of its integrable connection $\{\partial_i(\mathbf{x}) = B_i\mathbf{x}\}_{1 \le i \le m}$, and vice versa. Figure 1 illustrates such a relationship, and it also suggests reducing the problem of solving ∂-finite systems to that of solving fully integrable systems.

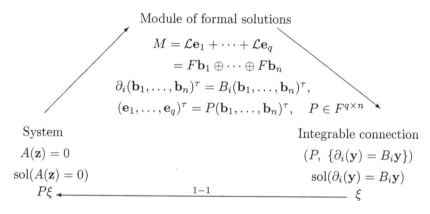

Fig. 1. Relationships among Systems, Modules and Solutions

4.4. *Fundamental matrices and Picard–Vessiot extensions*

Based on the discussion in Section 4.3, we generalize the notions and results of fundamental matrices and Picard–Vessiot extensions for ∂-finite systems.

Definition 4.2. Let $A(\mathbf{z}) = 0$ with $A \in \mathcal{S}^{p \times q}$ be a ∂-finite system, M be its module of formal solutions, $\{\mathbf{e}_1, \ldots, \mathbf{e}_q\}$ be a set of $\mathcal{L}$-generators of M and $\mathbf{b}_1, \ldots, \mathbf{b}_n$ be an F-basis of M such that $A(\mathbf{e}_1, \ldots, \mathbf{e}_q)^\tau = 0$ and $(\mathbf{e}_1, \ldots, \mathbf{e}_q)^\tau = P(\mathbf{b}_1, \ldots, \mathbf{b}_n)^\tau$ where $P \in F^{q \times n}$.

A $q \times n$ matrix V with entries in a Δ-extension E of F is called a *fundamental matrix* for $A(\mathbf{z}) = 0$ if $V = PU$ where $U \in E^{n \times n}$ is a fundamental matrix of the integrable connection of $A(\mathbf{z}) = 0$ with respect to $\mathbf{b}_1, \ldots, \mathbf{b}_n$.

A Picard–Vessiot ring for an integrable connection of $A(\mathbf{z}) = 0$ is called a *Picard–Vessiot ring* for $A(\mathbf{z}) = 0$.

As a consequence of Theorem 3.1, we have

Theorem 4.3. *Every ∂-finite system $A(\mathbf{z}) = 0$ over F has a Picard–Vessiot ring E. If F has characteristic 0 and C_F is algebraically closed, then $C_E = C_F$.*

Assume that F has characteristic 0 with an algebraically closed field of constants. If E is a Picard–Vessiot ring for the system $A(\mathbf{z}) = 0$ then the dimension of $\mathrm{sol}_E(A(\mathbf{z}) = 0)$ as a C_F-vector space equals the linear dimension of $A(\mathbf{z}) = 0$, whenever the latter is finite.

Example 4.2. Let F, δ_x, σ_k be as in Example 3.1, and

$$\mathcal{A} := \{\, \delta_x(\mathbf{z}) = A_x \mathbf{z}, \quad \sigma_k(\mathbf{z}) = A_k \mathbf{z} \,\}$$

where

$$A_x = \begin{pmatrix} \frac{x+1}{x} & \frac{k(x+1-k)}{x^2(k-1)} & -\frac{k(x+1-k)}{x^2(k-1)} \\ x+1 & \frac{xk-k^2+2x^2+kx^2+k-1}{x(k-1)} & -\frac{xk-k^2+2x^2+kx^2}{x(k-1)} \\ x+1 & \frac{xk+2x^2+kx^2-2k^2+k}{x(k-1)} & -\frac{xk+2x^2+kx^2-2k^2+1}{x(k-1)} \end{pmatrix},$$

$$A_k = \begin{pmatrix} \frac{k+1}{k} & \frac{k+1-xk-x}{x(k-1)} & \frac{xk+x-k-1}{x(k-1)} \\ \frac{x(k+1)}{k} & \frac{1-2x+k-xk+x^3}{k-1} & \frac{2x+xk-x^3-k-1}{k-1} \\ \frac{x(k+1)}{k} & \frac{1-2xk-2x+k+x^3}{k-1} & \frac{2xk+2x-k-x^3-1}{k-1} \end{pmatrix},$$

and $\mathbf{z} = (z_1, z_2, z_3)^\tau$. Note that A_k is singular, so $\mathcal{A}$ is integrable but not fully integrable. We will show in Example 5.2 that all solutions of $\mathcal{A}$ can be found by a change of variable $\mathbf{z} = P\mathbf{y}$ where

$$P = \begin{pmatrix} 1 & 0 \\ 0 & 1 \\ \frac{x(k-1)}{x^2-1} & \frac{x^2-k}{x^2-1} \end{pmatrix}$$

and $\mathbf{y}$ is a solution of the fully integrable system

$$\mathcal{B} : \{\delta_x(\mathbf{y}) = B_x \mathbf{y}, \ \sigma_k(\mathbf{y}) = B_k \mathbf{y}\}$$

with

$$B_x = \begin{pmatrix} \frac{-x+x^3-1+x^2-xk-k+k^2}{x(x^2-1)} & \frac{k(x+1-k)}{x^2(x^2-1)} \\ \frac{-x-xk+x^3-1-x^2+k^2-kx^2}{x^2-1} & \frac{-k^2+xk+kx^2+3x^2-1}{x(x^2-1)} \end{pmatrix}$$

and

$$B_k = \begin{pmatrix} \frac{xk+x+k^2+2k+1}{k(x+1)} & -\frac{k+1}{x(x+1)} \\ -\frac{(kx^2-x-k^2-2k-1)x}{k(x+1)} & \frac{x^2+x-1-k}{x+1} \end{pmatrix}.$$

So it suffices to compute a Picard–Vessiot extension of $\mathcal{B}$. The same method to construct a fundamental matrix for the system in Example 3.1 yields a fundamental matrix for $\mathcal{B}$:

$$U = \begin{pmatrix} xke^x & -kx^k \\ kx^2e^x & (x^2-k-1)x^{k+1} \end{pmatrix},$$

hence PU is a fundamental matrix for $\mathcal{A}$. In addition, a Picard–Vessiot ring $\mathbb{C}(x,k)[e^x, e^{-x}, x^k, x^{-k}]$ for $\mathcal{B}$ is a Picard–Vessiot ring for $\mathcal{A}$.

5. Computing linear dimension

We now describe how to compute linear dimension for a given linear functional system $A(\mathbf{z}) = 0$.

Let N be a submodule of a free $\mathcal{L}$-module $\mathcal{L}^q$ with a finite set of generators. One can compute a Gröbner basis of N over $\mathcal{L}$ (see Refs. 26 and [24, Chapter 3]), which gives rise to an F-basis of $\mathcal{L}^q/N$. Thus, one can determine whether a linear functional system is ∂-finite, and construct an F-basis of its module of formal solutions.

The following proposition indicates that the same goal may be achieved by Gröbner basis computation over Ore algebra $\mathcal{S}$ (see Ref. 8) if $\mathrm{coker}_{\mathcal{S}}(A)$ has finite dimension over F. Notice that the linear dimension of $A(\mathbf{z}) = 0$ never exceeds $\dim_F \mathrm{coker}_{\mathcal{S}}(A)$ by Remark 4.2.

Proposition 5.1. *Let N be a left submodule of $\mathcal{S}^{1 \times n}$ such that $\mathcal{S}^{1 \times n}/N$ has finite dimension over F and $\overline{N}$ the submodule generated by N in $\mathcal{L}^{1 \times n}$. Then $\mathcal{S}^{1 \times n}/(\overline{N} \cap \mathcal{S}^{1 \times n})$ and $\mathcal{L}^{1 \times n}/\overline{N}$ are isomorphic as F-vector spaces.*

The proof to Proposition 5.1 can be found in [24, Proposition 2.4.6].
Another useful fact is described in the next proposition whose proof is found in [24, Proposition 2.4.11(ii)].

Proposition 5.2. *Suppose that $\mathrm{coker}_{\mathcal{S}}(A)$ with $A \in \mathcal{S}^{p \times q}$ has a finite F-basis $\mathbf{f}_1, \ldots, \mathbf{f}_d$ and $\partial_i(\mathbf{f}_1, \ldots, \mathbf{f}_d)^\tau = D_i(\mathbf{f}_1, \ldots, \mathbf{f}_d)^\tau$ where $D_i \in F^{d \times d}$. Let D be the stacking of the blocks $(\partial_i \cdot \mathbf{1}_d - D_i)$. Then $\mathrm{coker}_{\mathcal{S}}(A) \cong_{\mathcal{S}} \mathrm{coker}_{\mathcal{S}}(D)$ and $\mathrm{coker}_{\mathcal{L}}(A) \cong_{\mathcal{L}} \mathrm{coker}_{\mathcal{L}}(D)$.*

The above proposition reveals that, to compute linear dimension of a system $A(\mathbf{z}) = 0$ such that $\mathrm{coker}_{\mathcal{S}}(A)$ has finite dimension over F, it suffices to compute linear dimension of the (integrable) system $\{\partial_i(\mathbf{y}) = D_i\mathbf{y}\}_{1 \leq i \leq m}$ where $\mathbf{y} = (y_1, \ldots, y_d)^\tau$.

In the situation described in Proposition 5.2, let H_0 be the set of row vectors of D, N_0 the submodule generated by H_0 over $\mathcal{S}$, and N the submodule generated by H_0 over $\mathcal{L}$. We proceed as follows to compute a Gröbner basis of N over $\mathcal{L}$, which gives rise to an F-basis of $\mathcal{L}^d/N$, the module of formal solutions of $\{\partial_i(\mathbf{y}) = D_i\mathbf{y}\}_{1 \leq i \leq m}$.

Compute the rank of D_j for each j with $\ell + 1 \leq j \leq m$. If each D_j has full rank, then $\{\partial_i(\mathbf{y}) = D_i\mathbf{y}\}_{1 \leq i \leq m}$ is fully integrable and so H_0 is already a Gröbner basis of N over $\mathcal{L}$ (see [24, Example 3.2.4]). Otherwise, the row vectors of some D_j satisfy certain linear relations over F, which, together with $\partial_j(\mathbf{y}) = D_j\mathbf{y}$, implies F-linear relations among $\partial_j(y_1), \ldots \partial_j(y_d)$. Assume that one of the relations is

$$f_1 \partial_j(y_1) + \cdots + f_d \partial_j(y_d) = 0$$

where $j \in \{\ell + 1, \ldots, m\}$ and $f_1, \ldots f_d \in F$, not all zero. Applying ∂_j^{-1} to it yields

$$\sigma_j^{-1}(f_1)\, y_1 + \cdots + \sigma_j^{-1}(f_d)\, y_d = 0. \tag{4}$$

Hence the vector $(\sigma_j^{-1}(f_1), \ldots, \sigma_j^{-1}(f_d))$ belongs to the $\mathcal{L}$-submodule N, but it does not belong to N_0. Adding to H_0 the new vectors obtained from the linear relations of the form (4), we have a new set H_1 of generators for N. Now we compute a Gröbner basis of N over $\mathcal{L}$ using H_1. By Lemmas 2.5.1 and 2.5.2 in Ref. 24, such a basis can be computed by merely rank computation, Gaussian elimination and the "Reduce All" trick. It is unnecessary to form any S-polynomials. This simplification is due to the integrability of $\{\partial_i(\mathbf{y}) = D_i\mathbf{y}\}_{1 \leq i \leq m}$. A detailed description of this process is formulated as an algorithm named LinearReduction in [24, Section 2.5]. We now give some examples to illustrate how to compute linear dimensions.

Example 5.1. Let $A_1, \ldots, A_m$ be in $F^{n \times n}$ and

$$A = \begin{pmatrix} \partial_1 \cdot \mathbf{1}_n - A_1 \\ \vdots \\ \partial_m \cdot \mathbf{1}_n - A_m \end{pmatrix} \in \mathcal{S}^{mn \times n}.$$

The system $A(\mathbf{z}) = 0$ corresponds to the system $\{\partial_i(\mathbf{z}) = A_i\mathbf{z}\}_{1 \leq i \leq m}$. Let M be the module of formal solutions of $A(\mathbf{z}) = 0$, and $\mathbf{e}_1, \ldots, \mathbf{e}_n$ be the respective images of $\mathbf{e}_{1n}, \ldots, \mathbf{e}_{nn}$ in M. For $\mathbf{e} := (\mathbf{e}_1, \ldots, \mathbf{e}_n)^\tau \in M^n$, we have $A(\mathbf{e}) = 0$ or $\partial_i(\mathbf{e}) = A_i\mathbf{e}$ for each i. Since the entries of A_i are in F, $\partial_i(\mathbf{e}_j) \in \sum_{s=1}^n F\mathbf{e}_s$ for all i, j, thus $\mathcal{L}\mathbf{e}_j \subseteq \sum_{s=1}^n F\mathbf{e}_s$ for all j.

So $M = \sum_{s=1}^{n} \mathcal{L}e_s = \sum_{s=1}^{n} F e_s$. In particular, $\dim_F M \leq n$. Observe that $\dim_F M = n$ if and only if the system $\{\partial_i(\mathbf{z}) = A_i \mathbf{z}\}_{1 \leq i \leq m}$ is fully integrable (see [24, Proposition 2.4.9]).

Example 5.2. Let F, δ_x, σ_k and the system $\mathcal{A}$ be given in Example 4.2. We follow the idea of the algorithm LinearReduction to compute linear dimension of $\mathcal{A}$. Note that A_k is singular. Solve the linear system $(v_1, v_2, v_3) A_k = 0$ in v_1, v_2, v_3. A nontrivial solution of this system yields

$$\sigma_k(z_3) = \frac{xk}{x^2 - 1}\sigma_k(z_1) + \frac{x^2 - k - 1}{x^2 - 1}\sigma_k(z_2).$$

By applying σ_k^{-1}, we get

$$z_3 = \frac{x(k-1)}{x^2 - 1}z_1 + \frac{x^2 - k}{x^2 - 1}z_2,$$

thus

$$\begin{pmatrix} z_1 \\ z_2 \\ z_3 \end{pmatrix} = \underbrace{\begin{pmatrix} 1 & 0 \\ 0 & 1 \\ \frac{x(k-1)}{x^2-1} & \frac{x^2-k}{x^2-1} \end{pmatrix}}_{P} \begin{pmatrix} z_1 \\ z_2 \end{pmatrix}.$$

Substitute this relation into $\mathcal{A}$, we get $\delta_x(z_1, z_2)^\tau = B_x(z_1, z_2)^\tau$ and $\sigma_k(z_1, z_2)^\tau = B_k(z_1, z_2)^\tau$ where

$$B_x = \begin{pmatrix} \frac{-x+x^3-1+x^2-xk-k+k^2}{x(x^2-1)} & \frac{k(x+1-k)}{x^2(x^2-1)} \\ \frac{-x-xk+x^3-1-x^2+k^2-kx^2}{x^2-1} & \frac{-k^2+xk+kx^2+3x^2-1}{x(x^2-1)} \end{pmatrix}$$

and

$$B_k = \begin{pmatrix} \frac{xk+x+k^2+2k+1}{k(x+1)} & -\frac{k+1}{x(x+1)} \\ -\frac{(kx^2-x-k^2-2k-1)x}{k(x+1)} & \frac{x^2+x-1-k}{x+1} \end{pmatrix}.$$

A straightforward calculation verifies that the first-order system $\mathcal{B}$ given by B_x and B_k is fully integrable, so $\mathcal{B}$ has linear dimension two by Example 5.1. According to the algorithm LinearReduction, all the solutions $\mathbf{z}$ of $\mathcal{A}$ can be obtained from the solutions $\mathbf{y}$ of $\mathcal{B}$ via a change of variable $\mathbf{z} = P\mathbf{y}$, and the modules of formal solutions for $\mathcal{A}$ and for $\mathcal{B}$ are the same. Hence $\mathcal{A}$ has linear dimension two. $\qquad\square$

The following example shows that there are ∂-finite systems whose $\mathcal{S}$-cokernels are infinite-dimensional over F.

Example 5.3. Let $L_1 = \partial_1\partial_2(\partial_1 + 1)$, $L_2 = \partial_1\partial_2(\partial_2 + 1)$, $A = (L_1, L_2)^\tau$, J the ideal in $\mathcal{S}$ generated by L_1 and L_2, and $M = \mathrm{coker}_{\mathcal{L}}(A)$. Since ∂_1 and ∂_2 are invertible in $\mathcal{L}$, then $M = \mathcal{L}/(\mathcal{L}(\partial_1 + 1) + \mathcal{L}(\partial_2 + 1))$, thus $\dim_F M = 1$. However, $\mathcal{S}/J$ is infinite-dimensional over F.

We now conclude how to determine whether a linear functional system is ∂-finite. As seen in Examples 5.1 and 5.2, when the system is given as an integrable system, we have a set of generators of M over F, so computing $\dim_F M$ can be done by linear algebra. When $A(\mathbf{z}) = 0$ is given by a finite-rank ideal in $\mathcal{S}$, Proposition 5.1 shows that either $M = 0$ (if the ideal contains a monomial in $\partial_{\ell+1}, \ldots, \partial_m$) or an F-basis of M can be computed via Gröbner bases of ideals in $\mathcal{S}$. There are algorithms and implementations for this task.[7,8] For a more general matrix $A \in \mathcal{S}^{p \times q}$, one can use the Gröbner basis technique developed in [24, Chapter 3] for computing F-bases of $\mathcal{L}$-modules. However, to compute the linear dimension of $A(\mathbf{z}) = 0$ for which $\mathrm{coker}_{\mathcal{S}}(A)$ is finite-dimensional it suffices to compute the linear dimension of an integrable system according to Proposition 5.2. The algorithm LinearReduction supplies a tool for the latter task. Therefore, Gröbner basis techniques in $\mathcal{L}$ are necessary only when $\mathrm{coker}_{\mathcal{S}}(A)$ is infinite-dimensional over F.

6. Factorization of Laurent–Ore modules

The work of this section is motivated by the algorithm FactorWithSpecifiedLeaders in Refs. 12,13, where the idea of associated equations is extended to factor linear partial differential equations with finite-dimensional solution spaces. In terms of modules over an Ore algebra $\mathcal{S} = F[\partial_1, \ldots, \partial_m]$ where $\ell = m$, the problem solved by their algorithm can be formulated as follows: *given a submodule N of $\mathcal{S}^n$ such that $M = \mathcal{S}^n/N$ is finite-dimensional over the field F, finds all submodules of $\mathcal{S}^n$ that contain N.* Such a submodule is called a factor of N since all its solutions are solutions of N. In their algorithm a factor is represented by a Gröbner basis with respect to a pre-chosen monomial order. Observe that, for a (right) factor of a given order, there is only one possibility for its leading derivative in the ordinary case, whereas, there are many possibilities in the partial case. Due to this complication, the algorithm has to check every possibility to compute all the factors of a given order. In this ideal-theoretic approach the quotient module M does not come into play.

In the module-theoretic approach to be described in this section, we compute all submodules of the above quotient module M, and then recover the "factors"of N in the sense of Refs. 12,13 via the canonical map from $\mathcal{S}^n$ to M. As all submodules of M are represented by linear bases over F, the problem of guessing leading derivatives goes away. The same idea carries over to Laurent–Ore modules and results in a factorization algorithm for ∂-finite systems.

6.1. *Constructions with modules over Laurent–Ore algebras*

Given a ring R, we first review some notions of reducibility of R-modules defined in Ref. 19.

An R-module M is *reducible* if M has a submodule other than 0 and M. Otherwise, M is *irreducible* or *simple*. An R-module M is *completely reducible* or *semisimple* if for every submodule N_1 there exists a submodule N_2 such that $M = N_1 \oplus N_2$. Note that an irreducible module is completely reducible as well. An R-module M is *decomposable* if M can be written as $N_1 \oplus N_2$ where N_1 and N_2 are nontrivial submodules of M. Otherwise, M is *indecomposable*. Clearly, an R-module M is reducible if it is decomposable, and M is irreducible when it is both indecomposable and completely reducible. By *factoring* an R-module, we mean finding its R-submodules.

As before, let F be a Δ-field with C the field of constants, $\mathcal{S} = F[\partial_1, \ldots, \partial_m]$ and $\mathcal{L} = F[\partial_1, \ldots, \partial_m, \partial_{\ell+1}^{-1}, \ldots, \partial_m^{-1}]$ be the corresponding Ore algebra and Laurent–Ore algebra, respectively. In the sequel, unless otherwise specified, F has characteristic 0 and C is algebraically closed.

Clearly, ordinary and partial differential modules in Ref. 22 are special cases of $\mathcal{L}$-modules. The constructions in [22, §2.2] can be carried on $\mathcal{L}$-modules in a similar way.

Let M be an $\mathcal{L}$-module and N a submodule of M. The F-vector space M/N with the induced actions:

$$\partial_i(\mathbf{w} + N) = \partial_i(\mathbf{w}) + N \quad \text{and} \quad \partial_j^{-1}(\mathbf{w} + N) = \partial_j^{-1}(\mathbf{w}) + N$$

for $1 \leq i \leq m$ and $\ell + 1 \leq j \leq m$, is the *quotient module*.

The *direct sum* of two $\mathcal{L}$-modules M_1 and M_2 is $M_1 \oplus M_2$ equipped with the actions:

$$\partial_i(\mathbf{w}_1 + \mathbf{w}_2) = \partial_i(\mathbf{w}_1) + \partial_i(\mathbf{w}_2) \quad \text{and} \quad \partial_j^{-1}(\mathbf{w}_1 + \mathbf{w}_2) = \partial_j^{-1}(\mathbf{w}_1) + \partial_j^{-1}(\mathbf{w}_2)$$

for $1 \leq i \leq m$ and $\ell + 1 \leq j \leq m$.

The *tensor product* $M_1 \otimes M_2$ of two $\mathcal{L}$-modules M_1 and M_2 is $M_1 \otimes_F M_2$

equipped with the actions:

$$\partial_i(\mathbf{w}_1 \otimes \mathbf{w}_2) = \partial_i(\mathbf{w}_1) \otimes \mathbf{w}_2 + \mathbf{w}_1 \otimes \partial_i(\mathbf{w}_2) \quad \text{for } i \leq \ell,$$

$$\partial_j^\nu(\mathbf{w}_1 \otimes \mathbf{w}_2) = \partial_j^\nu(\mathbf{w}_1) \otimes \partial_j^\nu(\mathbf{w}_2) \quad \text{for } j > \ell \text{ and } \nu \in \{-1, 1\}.$$

The d-th *exterior power* $\wedge^d M$ of an $\mathcal{L}$-module M is the F-vector space $\wedge_F^d M$ provided with the actions given by the formulas

$$\partial_i(\mathbf{w}_1 \wedge \cdots \wedge \mathbf{w}_d) = \sum_{s=1}^d \mathbf{w}_1 \wedge \cdots \wedge (\partial_i \mathbf{w}_s) \wedge \cdots \wedge \mathbf{w}_d,$$

$$\partial_j^\nu(\mathbf{w}_1 \wedge \cdots \wedge \mathbf{w}_d) = \partial_j^\nu(\mathbf{w}_1) \wedge \cdots \wedge \partial_j^\nu(\mathbf{w}_d),$$

for $i \leq \ell$, $j > \ell$ and $\nu \in \{-1, 1\}$.

Exterior powers of Laurent–Ore modules play an important role in the next section.

6.2. *A module-theoretic approach to factorization*

We now describe an idea on factoring Laurent–Ore modules.

Recall that a *decomposable*[16] element $\mathbf{w} \in \wedge^d M$ is an exterior product of d elements in M, *i.e.*, $\mathbf{w} = \mathbf{w}_1 \wedge \cdots \wedge \mathbf{w}_d$.

The following theorem generalizes Lemma 10 in Ref. 9 or the corresponding statement in [22, page 111]:

Theorem 6.1. *A Laurent–Ore module M has a d-dimensional submodule if and only if $\wedge^d M$ has a one-dimensional submodule generated by a decomposable element.*

Remark that the operators ∂_j^{-1} are indispensable in the proof of Theorem 6.1 (see also [24, Theorem 4.3.1]), and this proof yields a correspondence between d-dimensional submodules and one-dimensional submodules generated by decomposable elements: if a d-dimensional submodule of M has an F-basis $\mathbf{v}_1, \ldots, \mathbf{v}_d$, then the linear subspace generated by $\mathbf{v}_1 \wedge \cdots \wedge \mathbf{v}_d$ in $\wedge^d M$ is a one-dimensional submodule; conversely, if a one-dimensional submodule of $\wedge^d M$ is generated by a decomposable element $\mathbf{v}_1 \wedge \cdots \wedge \mathbf{v}_d$, then the F-linear subspace generated by $\mathbf{v}_1, \ldots, \mathbf{v}_d$ in M is a d-dimensional submodule.

Let M be a Laurent–Ore module with an F-basis $\{\mathbf{e}_1, \ldots, \mathbf{e}_n\}$ and set $g = \binom{n}{d}$. Then the module $\wedge^d M$ has an F-basis $\{\mathbf{f}_1, \ldots, \mathbf{f}_g\}$. Let

$$\mathbf{e} = (\mathbf{e}_1, \ldots, \mathbf{e}_n) \quad \text{and} \quad \mathbf{f} = (\mathbf{f}_1, \ldots, \mathbf{f}_g).$$

By Theorem 6.1, the problem of finding d-dimensional submodules of M is converted into that of finding one-dimensional submodules of $\wedge^d M$ whose generators are decomposable, and thus the factorization problem is reduced to two "subproblems": finding one-dimensional submodules and deciding the decomposability of their generators.

The first subproblem can be solved by a recursive method[14] for determining one-dimensional submodules of a Laurent–Ore module. Applying the method to $\wedge^d M$ yields several finite subsets $S_1, \ldots, S_t \subset F^g$ with the following properties:

(a) The elements of each S_k are C-linearly independent.

(b) A one-dimensional F-space of $\wedge^d M$ is a submodule if and only if it is an F-space generated by the product of $\mathbf{f}$ and a nontrivial C-linear combination of elements of some S_k.

Now we deal with the second subproblem. Let S be one of the S_k with q elements, and $\mathbf{w}$ be the product of $\mathbf{f}$ and a C-linear combination of elements of S, in which the coefficients are unspecified constants $c_1, \ldots, c_q$. Consider the map $\phi_{\mathbf{w}} : M \to \wedge^{d+1} M$ given by $\mathbf{v} \mapsto \mathbf{v} \wedge \mathbf{w}$. By Theorem 1.1 in [16, Chapter 4] and the proof of Theorem 6.1, $\mathbf{w}$ is decomposable if and only if $\ker(\phi_{\mathbf{w}})$ is of dimension d. The latter is equivalent to the condition that the matrix P of $\phi_{\mathbf{w}}$ has rank $(n - d)$. Hence, testing the decomposability of $\mathbf{w}$ amounts to a rank computation of P, i.e., identifying the unspecified constants $c_1, \ldots, c_q$ such that the rank of P is $(n-d)$. This further amounts to solving several systems consisting of homogeneous polynomial equations and inequations in $c_1, \ldots, c_q$ over F. Using a linear basis of F over C, we can translate every such system into finitely many subsystems over C. Each subsystem has two portions: a set of polynomial equations and an inequation. If none of the subsystems has a solution, then the product of $\mathbf{f}$ and any C-linear combination of elements of S is not decomposable and thus does not lead to any d-dimensional submodule of M. Otherwise, substitute a solution into the matrix P, and compute a basis $\mathbf{r}_1, \ldots, \mathbf{r}_d$ of the rational kernel of P where $\mathbf{r}_j \in F^n$. Set $\mathbf{u}_j = \mathbf{er}_j$ for $j = 1, \ldots, d$. Then $\oplus_{j=1}^d F\mathbf{u}_j$ is a d-dimensional submodule of M.

A few words need to be said about those subsystems derived from the rank condition for P, since they may have infinitely many solutions after dehomogenization. We require that the substitution of any solution of a subsystem into P not only yields the required rank for P, but also makes a fixed $(n-d) \times (n-d)$ minor nonzero. An $(n-d) \times (n-d)$ minor may correspond to several subsystems. This requirement can always be fulfilled, and will help us describe all d-submodules of M by a finite amount of information. We proceed as follows. Let T be such a subsystem. Using the nonzero minor corresponding to T and Cramer's rule, we may find a basis $\mathbf{r}_1, \ldots, \mathbf{r}_d$ of the rational kernel of P where the entries of the $\mathbf{r}_j$ are in $F(c_1, \ldots, c_q)$ and their denominators divide the given minor. Set $\mathbf{u}_j = \mathbf{er}_j$ for $j = 1, \ldots, d$. Then $\oplus_{j=1}^d F\mathbf{u}_j$ represents all d-dimensional submodules obtained by substituting solutions of T for $c_1, \ldots, c_q$ into $\mathbf{u}_1, \ldots, \mathbf{u}_d$. Note that we may check the set of solutions of T by techniques from computational algebraic geometry. These considerations lead to a method for computing all submodules of M, which is described stepwise in [24, Section 4.4].

Remark 6.1. The above representation for d-dimensional submodules of M is rather naive and has a lot of redundancy. To have more concise representations, one would partition d-dimensional submodules with respect to module isomorphism, and generalize the techniques given in [22, page 112].

We present two examples for factoring Laurent–Ore modules.

Example 6.1. [Legendre's system] Let F, δ_x, σ_k be as given in Example 3.1 and $\mathcal{L} = F[\partial_x, \partial_k, \partial_k^{-1}]$ be the Laurent–Ore algebra. A Gröbner basis of the ideal generated by the Legendre's system (1), is

$$g_1 = xk + x + (x^2 - 1)\partial_x - (k+1)\partial_k \quad \text{and} \quad g_2 = k + 1 + (k+2)\partial_k^2 - (2xk + 3x)\partial_k.$$

Let $A = (g_1, g_2)^\tau \in \mathcal{L}^{2 \times 1}$, $M = \mathcal{L}/(\mathcal{L}g_1 + \mathcal{L}g_2)$ and $\mathbf{e}_1, \mathbf{e}_2$ be the images of 1 and ∂_k in M, respectively. Then $\mathbf{e}_1, \mathbf{e}_2$ form a basis of M over F and, in addition,

$$\partial_x \begin{pmatrix} \mathbf{e}_1 \\ \mathbf{e}_2 \end{pmatrix} = \begin{pmatrix} \frac{-xk-x}{x^2-1} & \frac{k+1}{x^2-1} \\ \frac{-k-1}{x^2-1} & \frac{xk+x}{x^2-1} \end{pmatrix} \begin{pmatrix} \mathbf{e}_1 \\ \mathbf{e}_2 \end{pmatrix}, \quad \partial_k \begin{pmatrix} \mathbf{e}_1 \\ \mathbf{e}_2 \end{pmatrix} = \begin{pmatrix} 0 & 1 \\ \frac{-k-1}{k+2} & \frac{2xk+3x}{k+2} \end{pmatrix} \begin{pmatrix} \mathbf{e}_1 \\ \mathbf{e}_2 \end{pmatrix}.$$

Apply the algorithm in Ref. 14, we find that M has no one-dimensional submodules, so M is irreducible.

Example 6.2. Let F, δ_x, σ_k be as in Example 3.1 and $\mathcal{L} = F[\partial_x, \partial_k, \partial_k^{-1}]$ the Laurent–Ore algebra. Let M be an $\mathcal{L}$-module with an F-basis $\{\mathbf{e}_1, \mathbf{e}_2, \mathbf{e}_3, \mathbf{e}_4\}$ satisfying $\partial_x(\mathbf{e}_1, \mathbf{e}_2, \mathbf{e}_3, \mathbf{e}_4)^\tau = A_x(\mathbf{e}_1, \mathbf{e}_2, \mathbf{e}_3, \mathbf{e}_4)^\tau$ and $\partial_k(\mathbf{e}_1, \mathbf{e}_2, \mathbf{e}_3, \mathbf{e}_4)^\tau = A_k(\mathbf{e}_1, \mathbf{e}_2, \mathbf{e}_3, \mathbf{e}_4)^\tau$ where

$$A_x = \begin{pmatrix} 0 & 1 & 0 & 0 \\ a_1 & a_2 & 0 & 0 \\ 0 & 0 & 0 & 1 \\ 0 & 0 & a_3 & a_4 \end{pmatrix}$$

with

$$a_1 = \frac{-x^3 - x^2k + 2x^2 + xk + k^2x + k^2 + k^3}{x^2(-x+k)}, \quad a_2 = \frac{2(x^2 - x - k^2)}{(x-k)x},$$

$$a_3 = \frac{-x^3 - x^2k + x^2 + 3xk + 2x + k^2x + 4k^2 + 5k + 2 + k^3}{x^2(-x+k+1)},$$

$$a_4 = -\frac{2(-x^2 + x + k^2 + 2k + 1)}{(-x+k+1)x},$$

and

$$A_k = \begin{pmatrix} 0 & 0 & 1 & 0 \\ 0 & 0 & 0 & 1 \\ -\frac{(x-k)x^2}{x-k-2} & 0 & \frac{2x(x-k-1)}{x-k-2} & 0 \\ -\frac{2x(x^2-2xk-3x+k^2+2k)}{(x-k-2)^2} & -\frac{(x-k)x^2}{x-k-2} & \frac{2(x^2-2xk-4x+k^2+3k+2)}{(x-k-2)^2} & \frac{2x(x-k-1)}{x-k-2} \end{pmatrix}.$$

Let us compute all two-dimensional submodules of M. Clearly,

$$\mathbf{f}_1 = \mathbf{e}_1 \wedge \mathbf{e}_2, \ \ \mathbf{f}_2 = \mathbf{e}_1 \wedge \mathbf{e}_3, \ \ \mathbf{f}_3 = \mathbf{e}_1 \wedge \mathbf{e}_4, \ \ \mathbf{f}_4 = \mathbf{e}_2 \wedge \mathbf{e}_3, \ \ \mathbf{f}_5 = \mathbf{e}_2 \wedge \mathbf{e}_4, \ \ \mathbf{f}_6 = \mathbf{e}_3 \wedge \mathbf{e}_4$$

form a basis of $\wedge^2 M$ over F. By the algorithm in Ref. 14, every one-dimensional submodule of $\wedge^2 M$ has a generator of the form

$$\mathbf{w} = (\mathbf{f}_1, \mathbf{f}_2, \mathbf{f}_3, \mathbf{f}_4, \mathbf{f}_5, \mathbf{f}_6) \left(\sum_{i=1}^{6} c_i \mathbf{v}_i \right)$$

where

$$\mathbf{v}_1 = \begin{pmatrix} -x^2 + 2xk - k^2 \\ \frac{2k^3+k^2-2r^2-r^3+r+k^4-2x^2k^2+3xk+3k^2x+4x^2k+x^4}{x^3} \\ \frac{x+2x^3+k^3+2xk+k^2-3x^2k-3x^2}{x^2} \\ \frac{-2xk+k+k^3+x^2k+2k^2-x-2k^2x}{x^2} \\ -\frac{k+k^2+x^2-x-2xk}{x} \\ -\frac{x^2-2xk-2x+k^2+2k+1}{x^2} \end{pmatrix},$$

and the other expressions $\mathbf{v}_2, \ldots, \mathbf{v}_6$ are quite big and are given in [24, pages 70–71].

It remains to determine the decomposability of $\mathbf{w}$. Consider the map $M \to \wedge^3 M$ given by $\mathbf{v} \mapsto \mathbf{v} \wedge \mathbf{w}$, whose matrix is some $P \in F^{4 \times 4}$ (we do not write down this matrix explicitly due to its big size). The matrix P has rank 2 if and only if all its 3×3 minors are zero and there exists a nonzero 2×2

minor. This yields four sets of solutions for the c_i:

$$\{ c_1 = c_1, \quad c_2 = c_2, \quad c_3 = 0, \quad c_4 = c_4, \quad c_5 = 0, \quad c_6 = 0 \},$$

$$\left\{ c_1 = c_1, \quad c_2 = c_2, \quad c_3 = c_3, \quad c_4 = -\frac{3c_3}{2}, \quad c_5 = 0, \quad c_6 = 0 \right\},$$

$$\left\{ c_1 = \frac{c_3(2c_4 + 3c_3 - 4c_5)}{4c_5}, \quad c_2 = c_2, \ c_3 = c_3, \ c_4 = c_4, \ c_5 = c_5, \ c_6 = 0 \right\},$$

$$\left\{ c_1 = c_1, c_2 = -\frac{8c_6c_5 - 2c_3c_4 - 3c_3^2 - 4c_6c_4 - 20c_3c_6 - 4c_1c_6 + 4c_1c_5 + 4c_3c_5 - 24c_6^2}{4c_6}, \right.$$

$$c_3 = c_3, \quad c_4 = c_4, \quad c_5 = c_5, \quad c_6 = c_6 \}.$$

Therefore M has two-dimensional submodules if and only if the c_i in $\mathbf{w}$ satisfy one of these four relations.

Substitute these four relations into P respectively and compute the corresponding F-bases for the rational kernel of P. Finally, we get all two-dimensional submodules:

$$N_i = \{ a_1 \mathbf{u}_{i,1} + a_2 \mathbf{u}_{i,2} \mid a_1, a_2 \in F \}, \quad i = 1, 2, 3, 4,$$

where

$$
\begin{aligned}
\mathbf{u}_{1,1} = {} & (2c_1 x^2 k + c_1 k^2 x + c_4 x^5 + c_1 x^2 + 2c_1 x^4 + c_2 x^2 k - 2x^3 c_2 k - 3x^3 c_1 k \\
& - 2x^4 c_4 k + x^2 c_2 k^2 + x k^3 c_1 + x^4 c_2 + c_4 x^3 k^2 - c_2 x^3 - 2c_4 x^4 - 3c_1 x^3 \\
& + c_4 x^3 + 2c_4 x^3 k)/(x(2c_1 x - c_1 x^2 + 2c_1 x k - 2c_2 k - c_2 k^2 + k^3 c_4 + 2c_2 x \\
& - c_2 x^2 + 2k^2 c_4 - 2c_1 k - c_1 k^2 + 2c_2 x k - 2k c_4 x - 2k^2 c_4 x + k c_4 x^2 + k c_4 \\
& - c_1 - c_2))e_1 - (c_4 x^3 + c_1 x^2 - k c_4 x^2 - c_4 x^2 - c_1 x k - c_1 x)(x - k)/(2c_1 x \\
& - c_1 x^2 + 2c_1 x k - 2c_2 k - c_2 k^2 + k^3 c_4 + 2c_2 x - c_2 x^2 + 2k^2 c_4 - 2c_1 k \\
& - c_1 k^2 + 2c_2 x k - 2k c_4 x - 2k^2 c_4 x + k c_4 x^2 + k c_4 - c_1 - c_2)e_2 + e_3,
\end{aligned}
$$

$$
\begin{aligned}
\mathbf{u}_{1,2} = {} & (c_1 x^4 - 2c_1 x^2 + c_4 x^5 + c_1 x - 2c_4 x^3 - c_1 x^3 - c_4 x^4 + c_2 x^3 - 2c_2 x^2 \\
& + 2c_4 x^2 + c_1 k^2 + 2c_1 k^3 + c_1 k^4 - 2c_2 x^2 k + 3c_1 k^2 x - 4c_1 x^2 k + 5k c_4 x^2 \\
& + k^2 c_4 x + c_2 x k + 3c_1 x k - 4c_4 x^3 k - 2c_4 x^3 k^2 - 2c_1 x^2 k^2 + 2c_4 x k^3 \\
& + 3c_4 x^2 k^2 + c_4 x k^4 + c_2 x k^2)/(x(2c_1 x - c_1 x^2 + 2c_1 x k - 2c_2 k - c_2 k^2 \\
& + k^3 c_4 + 2c_2 x - c_2 x^2 + 2k^2 c_4 - 2c_1 k - c_1 k^2 + 2c_2 x k - 2k c_4 x - 2k^2 c_4 x \\
& + k c_4 x^2 + k c_4 - c_1 - c_2))e_1 - (c_4 x^3 - c_2 x^2 + k c_4 x^2 + c_2 x k - 2k^2 c_4 x \\
& + c_2 x - 2x c_4 + c_1 x k - 4k c_4 x - c_1 - c_1 k^2 - 2c_1 k)(x - k)/((2c_1 x - c_1 x^2 \\
& + 2c_1 x k - 2c_2 k - c_2 k^2 + k^3 c_4 + 2c_2 x - c_2 x^2 + 2k^2 c_4 - 2c_1 k - c_1 k^2 \\
& + 2c_2 x k - 2k c_4 x - 2k^2 c_4 x + k c_4 x^2 + k c_4 - c_1 - c_2))e_2 + e_4.
\end{aligned}
$$

The submodules given by the last three solutions of the c_i's are more complicated (see [24, Example 4.4.1]).

6.3. *Eigenrings and decomposition of Laurent–Ore modules*

We discuss another approach to factoring Laurent–Ore modules, which is not based on the associated equations method. This method is first introduced in Ref. 20 to factor linear ordinary differential operators using eigenrings of the operators. Three algorithms are presented there for computing eigenrings. Significant improvements on these algorithms are described in Refs. 4,23. Although the eigenring method does not always factor reducible operators, it often yields factors quickly. This method has been generalized in Refs. 1,5 for systems of linear difference equations, and in Ref. 3 recently for systems of linear partial differential equations in positive characteristic. We will generalize this method for factoring Laurent–Ore modules.

Let R be an arbitrary ring and M be an R-module. Recall that $\mathrm{End}_R(M)$ is the set of all R-linear maps on M. Clearly, $\mathrm{End}_R(M)$ becomes a ring with the usual addition and the composition of maps adopted as the multiplication.

Definition 6.1. Let M be an R-module. A set of elements $\pi_1, \ldots, \pi_s$ of $\mathrm{End}_R(M)$ is called *a set of orthogonal idempotents* if they satisfy

$$\sum_{i=1}^{s} \pi_i = \mathbf{1} \quad \text{and} \quad \pi_i \pi_j = \mathbf{0} \quad \text{whenever } i \neq j, \tag{5}$$

where $\mathbf{1}$ and $\mathbf{0}$ are the identity map and the zero map on M, respectively.

Although it is not stated in Definition 6.1, the maps π_i are all idempotent. Indeed, the condition (5) implies that

$$\pi_i^2 = \sum_{j=1}^{s} \pi_i \pi_j = \pi_i \left(\sum_{j=1}^{s} \pi_j \right) = \pi_i, \quad \text{for each } i.$$

It is stated in Exercise 7 of [11, (Chapter 1, §1)] that

Proposition 6.1. *Let M be an R-module. If $\mathrm{End}_R(M)$ has a set of orthogonal idempotents $\pi_1, \ldots, \pi_s$ then $M = \oplus_{i=1}^{s} \pi_i(M)$. Conversely, if M can be written as a direct sum of submodules $M = N_1 \oplus \cdots \oplus N_s$ then $\{\pi_1, \ldots, \pi_s\}$ is a set of orthogonal idempotents of $\mathrm{End}_R(M)$ where π_i is the projection from M to N_i.*

For any R-module M, $\mathrm{End}_R(M)$ always has a set of orthogonal idempotents $\{\mathbf{0}, \mathbf{1}\}$, which is called the *trivial orthogonal idempotents* of $\mathrm{End}_R(M)$. As a direct consequence of Proposition 6.1, an R-module M is decomposable if and only if $\mathrm{End}_R(M)$ contains a nontrivial set of orthogonal idempotents.

Let F be a Δ-field with C the field of constants and $\mathcal{L} = F[\partial_1, \ldots, \partial_m, \partial_{\ell+1}^{-1}, \ldots, \partial_m^{-1}]$ the Laurent–Ore algebra over F. Here we do not assume that C is algebraically closed.

For an $\mathcal{L}$-module M, the endomorphism ring $\mathrm{End}_{\mathcal{L}}(M)$ is called the *eigenring* of M and denoted $\mathcal{E}(M)$. Then a map $\phi \in \mathrm{End}_F(M)$ belongs to $\mathcal{E}(M)$ if and only if ϕ commutes with the ∂_i and ∂_j^{-1} for all i, j with $1 \le i \le m$ and $\ell + 1 \le j \le m$. However, since M is an $\mathcal{L}$-module on which the ∂_j^{-1} act, the commutativity of ϕ with the ∂_j for $\ell + 1 \le j \le m$ implies $\partial_j \circ \phi \circ \partial_j^{-1}(\mathbf{w}) = \phi(\mathbf{w})$ and further $\phi \circ \partial_j^{-1}(\mathbf{w}) = \partial_j^{-1} \circ \phi(\mathbf{w})$ for $\mathbf{w} \in M$. Hence, $\phi \in \mathrm{End}_F(M)$ belongs to $\mathcal{E}(M)$ if and only if ϕ commutes with all the ∂_i for $1 \le i \le m$.

Let M be a Laurent–Ore module with an F-basis $\{\mathbf{e}_1, \ldots, \mathbf{e}_n\}$. Suppose that $\partial_i(\mathbf{e}_1, \ldots, \mathbf{e}_n)^\tau = B_i(\mathbf{e}_1, \ldots, \mathbf{e}_n)^\tau$ where $B_i \in F^{n \times n}$ for $1 \le i \le m$ and the B_j are invertible for $j > \ell$. In practice, the maps in $\mathcal{E}(M)$ can be interpreted in terms of the B_i. Let $\phi \in \mathrm{End}_F(M)$ and $P \in F^{n \times n}$ be its transformation matrix given by $(\phi(\mathbf{e}_1), \ldots, \phi(\mathbf{e}_n))^\tau = P(\mathbf{e}_1, \ldots, \mathbf{e}_n)^\tau$. Let $\mathbf{w} = \sum_{i=1}^n a_i \mathbf{e}_i \in M$ where $a_i \in F$. Then

$$\phi(\mathbf{w}) = \sum_{i=1}^n a_i \phi(\mathbf{e}_i) = (a_1, \ldots, a_n)(\phi(\mathbf{e}_1), \ldots, \phi(\mathbf{e}_n))^\tau$$
$$= (a_1, \ldots, a_n)P(\mathbf{e}_1, \ldots, \mathbf{e}_n)^\tau.$$

One can verify that the conditions $\partial_i(\phi(\mathbf{w})) = \phi(\partial_i(\mathbf{w}))$ hold for $\mathbf{w} \in M$ and $1 \le i \le m$ if and only if $\delta_i(P) = B_i P - P B_i$ for $i \le \ell$ and $\sigma_j(P) = B_j P B_j^{-1}$ for $j > \ell$. Hence the eigenring $\mathcal{E}(M)$ can be defined equivalently to be

$$\mathcal{E}(M) = \{P \in F^{n \times n} \mid \delta_i(P) = B_i P - P B_i, i \le \ell, \sigma_j(P) = B_j P B_j^{-1}, j > \ell\}. \tag{6}$$

Clearly, the identity matrix $\mathbf{1}_n \in \mathcal{E}(M)$ and $\mathcal{E}(M)$ is a C-subalgebra of $F^{n \times n}$ of dimension not greater than n^2. Moreover, $C \cdot \mathbf{1}_n \subseteq \mathcal{E}(M)$ where $C \cdot \mathbf{1}_n$ denotes the set of all matrices of the form $c \cdot \mathbf{1}_n$ where $c \in C$. As a natural generalization of the results in Ref. 22, [20, Proposition 2.13] or Ref. 2 for the case of linear ordinary differential equations, we have

Theorem 6.2. *Let M be an $\mathcal{L}$-module of dimension n. Then*

(i) *If $\mathcal{E}(M) \ne C \cdot \mathbf{1}_n$ then M is reducible.*
(ii) *If M is decomposable then $\mathcal{E}(M) \ne C \cdot \mathbf{1}_n$.*
(iii) *If M is completely reducible, then M is irreducible if and only if $\mathcal{E}(M) = C \cdot \mathbf{1}_n$.*

Given a Laurent–Ore module M of dimension n, we now use the formula (6) to compute $\mathcal{E}(M)$. Let $P \in \mathcal{E}(M)$ be a matrix of n^2 indeterminates z_{ij}. From (6), we get a system $\partial_i(\mathbf{z}) = A_i \mathbf{z}$ where $A_i \in F^{n^2 \times n^2}$ and $\mathbf{z} = (z_{11}, \ldots, z_{1n}, \ldots, z_{n1}, \ldots, z_{nn})^\tau$. This system is clearly ∂-finite, so its rational solutions can be found by a specialized version of the method in Ref. 14. A C-basis of all rational solutions of this system yields a C-basis $\{P_1, \ldots, P_r\}$ of all rational solutions of $\mathcal{E}(M)$. Without loss of generality, we assume that $P_1 = \mathbf{1}_n$. Therefore $\mathcal{E}(M) = \oplus_{i=1}^r C \cdot P_i$. If $r = 1$, then $\mathcal{E}(M)$ is trivial and M is indecomposable by Theorem 6.2 (ii). Otherwise, each eigenvalue λ of a nontrivial $P \in \mathcal{E}(M)$ will produce a submodule $\{\mathbf{w} \in M \mid P\mathbf{w} = \lambda\mathbf{w}\}$ of M. If $\mathcal{E}(M)$ has a set of nontrivial orthogonal idempotents $\pi_1, \ldots, \pi_s$, then we derive a decomposition $M = \pi_1(M) \oplus \cdots \oplus \pi_s(M)$. If M is furthermore completely reducible, a maximal decomposition of M can be obtained by applying the eigenring method recursively on the submodules in the above decomposition.

Example 6.3. Let F, δ_x, σ_k be as given in Example 3.1, $\mathcal{L} = F[\partial_x, \partial_k, \partial_k^{-1}]$ the Laurent–Ore algebra and M an $\mathcal{L}$-module of dimension two. Suppose that $\{\mathbf{e}_1, \mathbf{e}_2\}$ is a basis of M satisfying $\partial_x(\mathbf{e}_1, \mathbf{e}_2)^\tau = B_x(\mathbf{e}_1, \mathbf{e}_2)^\tau$ and $\partial_k(\mathbf{e}_1, \mathbf{e}_2)^\tau = B_k(\mathbf{e}_1, \mathbf{e}_2)^\tau$ where

$$B_x = \begin{pmatrix} 1 & 0 \\ 0 & 0 \end{pmatrix} \quad \text{and} \quad B_k = \begin{pmatrix} 1 & 0 \\ 0 & k \end{pmatrix}.$$

We now compute the eigenring of M. Let $P \in \mathcal{E}(M)$ be a 2×2 matrix with indeterminate entries z_{11}, z_{12}, z_{21} and z_{22}. The conditions

$$\delta_x(P) = B_x P - P B_x \quad \text{and} \quad \sigma_k(P) = B_k P B_k^{-1}$$

yields a system $\mathcal{A} : \{\delta_x(\mathbf{z}) = A_x \mathbf{z}, \sigma_k(\mathbf{z}) = A_k \mathbf{z}\}$ where $\mathbf{z} = (z_{11}, z_{12}, z_{21}, z_{22})^\tau$,

$$A_x = \begin{pmatrix} 0 & \frac{k\eta}{(x-k)(x-1)} & \frac{\eta+k-x}{x(x-k)k(x-1)} & 0 \\[2mm] -\frac{\eta+k-x}{x(x-k)k(x-1)} & -\frac{x^3-kx^2-2x+3k+kx}{x(x-k)(x-1)} & 0 & \frac{\eta+k-x}{x(x-k)k(x-1)} \\[2mm] -\frac{k\eta}{(x-k)(x-1)} & 0 & \frac{x^3-kx^2-2x+3k+kx}{x(x-k)(x-1)} & \frac{k\eta}{(x-k)(x-1)} \\[2mm] 0 & -\frac{k\eta}{(x-k)(x-1)} & -\frac{\eta+k-x}{x(x-k)k(x-1)} & 0 \end{pmatrix}$$

and

$$
A_k = \frac{1}{\gamma}
\begin{pmatrix}
\frac{\alpha\xi}{k} & -x\alpha\beta & -\frac{\beta\xi}{k^2} & \frac{x\beta^2}{k} \\
\frac{\alpha\beta}{k(k+1)} & \frac{\alpha^2}{k+1} & -\frac{\beta^2}{k^2(k+1)} & -\frac{\alpha\beta}{k(k+1)} \\
\frac{x(k+1)\beta\xi}{k} & -(k+1)x^2\beta^2 & \frac{(k+1)\xi}{k^2} & -\frac{x(k+1)\beta\xi}{k} \\
\frac{x\beta^2}{k} & x\alpha\beta & \frac{\beta\xi}{k^2} & \frac{\alpha\xi}{k}
\end{pmatrix},
$$

with $\eta = x^2 - kx + 2k - x$, $\alpha = k+1+kx^2-k^2x-x$, $\beta = k+1+kx-k^2-x$, $\gamma = (x-k)(x-k-1)(x-1)^2$ and $\xi = x^2 - 2kx - x + k^2$. All rational solutions of $\mathcal{A}$ are of the form

$$
\left(\frac{c_1 - c_2 x}{x-1}, \; \frac{c_2 - c_1}{k(x-1)}, \; \frac{(c_1 - c_2)xk}{x-1}, \; \frac{c_1 x - c_2}{x-1} \right), \quad \text{for } c_1, c_2 \in \mathbb{C}.
$$

So

$$
\mathcal{E}(M) = \left\{ \begin{pmatrix} \frac{c_1 - c_2 x}{x-1} & \frac{c_2 - c_1}{k(x-1)} \\ \frac{(c_1 - c_2)xk}{x-1} & \frac{c_2 - c_1 x}{x-1} \end{pmatrix}, \quad \text{for any } c_1, c_2 \in \mathbb{C} \right\}
$$

$$
= \mathbb{C} \begin{pmatrix} \frac{1}{x-1} & -\frac{1}{k(x-1)} \\ \frac{kx}{x-1} & -\frac{x}{x-1} \end{pmatrix} \oplus \mathbb{C} \begin{pmatrix} -\frac{x}{x-1} & \frac{1}{k(x-1)} \\ -\frac{kx}{x-1} & \frac{1}{x-1} \end{pmatrix}.
$$

Recall that the necessary condition for $\{P_1, \ldots, P_s\} \subset \mathcal{E}(M)$ being a set of orthogonal idempotents is that $P_i^2 = P_i$ for each i. Substitute

$$
P = \begin{pmatrix} \frac{c_1 - c_2 x}{x-1} & \frac{c_2 - c_1}{k(x-1)} \\ \frac{(c_1 - c_2)xk}{x-1} & \frac{c_2 - c_1 x}{x-1} \end{pmatrix}
$$

into the relation $P^2 = P$, we obtain three solutions:

$$
P_0 = \mathbf{1}_2, \quad P_1 = \begin{pmatrix} -\frac{1}{x-1} & \frac{1}{k(x-1)} \\ -\frac{kx}{x-1} & \frac{x}{x-1} \end{pmatrix}, \quad P_2 = \begin{pmatrix} \frac{x}{x-1} & -\frac{1}{k(x-1)} \\ \frac{kx}{x-1} & -\frac{1}{x-1} \end{pmatrix}.
$$

Among which, we find $P_1 P_2 = 0$ and $P_1 + P_2 = \mathbf{1}_2$. So $\{P_1, P_2\}$ is a set of nontrivial orthogonal idempotents of $\mathcal{E}(M)$. We have

$$
P_1(M) = \{P_1(\mathbf{w}) \mid \mathbf{w} \in M\}
$$
$$
= \{(a_1, a_2)P_1(\mathbf{e}_1, \mathbf{e}_2)^\tau \mid a_1, a_2 \in F\} = F \cdot \left(\mathbf{e}_1 - \frac{1}{k}\mathbf{e}_2 \right)
$$

and

$$
P_2(M) = \{P_2(\mathbf{w}) \mid \mathbf{w} \in M\}
$$
$$
= \{(a_1, a_2)P_2(\mathbf{e}_1, \mathbf{e}_2)^\tau \mid a_1, a_2 \in F\} = F \cdot \left(\mathbf{e}_1 - \frac{1}{kx}\mathbf{e}_2 \right).
$$

Therefore, $P_1(M) \oplus P_2(M)$ is a decomposition of M into two nontrivial submodules.

The eigenring method, however, may fail to find any factor of a Laurent–Ore module even this module is reducible. This happens when the module is not completely reducible and its eigenring is trivial.

7. Concluding remarks

In this paper we have discussed how to solve and factor ∂-finite systems. A key technique described here is to use the notion of modules of formal solutions to connect ∂-finite systems with fully integrable systems, while the latter systems are very similar to linear ordinary differential (difference) equations. This technique naturally gives rise to Picard–Vessiot extensions for ∂-finite systems. Since Picard–Vessiot extensions are a stepping-stone to introduce Galois groups, it would be interesting to extend (part of) the Galois theory for linear ordinary (difference) equations to ∂-finite systems. We presented some methods for determining linear dimension of a linear functional system. We also generalized Beke's method and the eigenring approach to factor Laurent–Ore modules. The work on factoring Laurent–Ore modules is however preliminary, because efficiency and applications of these two methods have not yet been considered.

References

1. M.A. Barkatou. Rational solutions of matrix difference equations: the problem of equivalence and factorization. In *Proc. ISSAC'99*, pages 277–282, Vancouver, Canada, 1999.
2. M.A. Barkatou. On the reduction of matrix pseudo-linear equations. Technical report, Rapport de Recherche de L'institut IMAG, Grenoble, France, 2001.
3. M.A. Barkatou, T. Cluzeau, and J.-A. Weil. Factoring partial differential systems in positive characteristic. In D. Wang, editor, *Differential Equations and Symbolic Computation*, chapter 13. Birkhauser, Basel Boston, 2005.
4. M.A. Barkatou and E. Pflügel. On the equivalence problem of linear differential systems and its application for factoring completely reducible systems. In *Proc. ISSAC'98*, Rostock, Germany, 1998.
5. R. Bomboy. *Réductibilité et Résolubilité des Équations aux Différences Finies*. PhD thesis, Université de Nice–Sophia Antipolis, Sophia Antipolis, France, 2001.
6. M. Bronstein, Z. Li, and M. Wu. Picard–Vessiot extensions for linear functional systems. In *Proc. ISSAC'2005*, Beijing, China, 2005.
7. F. Chyzak, A. Quadrat, and D. Robertz. OREMODULES: A symbolic package for the study of multidimensional linear systems. In *Proc. MTNS'2004*, Leuven, Belgium, 2004.
8. F. Chyzak and B. Salvy. Non-commutative elimination in Ore algebras proves multivariate identities. *J. Symbolic Computation*, 26(2):187–228, 1998.

9. E. Compoint and J.-A. Weil. Absolute reducibility of differential operators and Galois groups. *J. of Algebra*, 275(1):77–105, 2003.
10. G. Labahn and Z. Li. Hyperexponential solutions of finite-rank ideals in orthogonal Ore algebras. In *Proc. ISSAC'2004*, pages 213–220, Santander, Spain, 2004.
11. T. Y. Lam. *A First Course in Non-Commutative Rings, Graduate Texts in Mathematics*, volume 131. Springer–Verlag, 1991.
12. Z. Li, F. Schwarz, and S. Tsarev. Factoring zero-dimensional ideals of linear partial differential operators. In *Proc. ISSAC'2002*, pages 168–175, Lille, France, 2002.
13. Z. Li, F. Schwarz, and S. Tsarev. Factoring systems of linear PDE's with finite-dimensional solution spaces. *J. Symbolic Computation*, 36:443–471, 2003.
14. Z. Li, M. F. Singer, M. Wu, and D. Zheng. A recursive method for determining the one-dimensional submodules of Laurent–Ore modules. In *Proc. ISSAC'2006*, pages 220–227, Genova, Italy, 2006.
15. B. Malgrange. Motivations and introduction to the theory of *D*-modules. In E. Tournier, editor, *Computer Algebra and Differential Equations, volume 193 of LMS Lecture Note Series*, pages 3–20. Cambridge University Press, 1994.
16. M. Marcus. *Finite Dimensional Multilinear Algebra, Part II*. Marcel Dekker, New York, 1975.
17. F. Pauer and A. Unterkircher. Gröbner bases for ideals in Laurent polynomial rings and their applications to systems of difference equations. In *Proc. AAECC'99*, volume 9, pages 271–291, Hawaii, USA, 1999.
18. J. J. Rotman. *An Introduction to Homological Algebra*. Academic Press, New York, 1979.
19. L. H. Rowen. *Ring Theory, I*. San Diego, CA:Academic Press, 1988.
20. M. F. Singer. Testing reducibility of linear differential operators: a group theoretic perspective. *J. Appl. Alg. in Eng. Comm. and Compu.*, 7(2):77–104, 1996.
21. M. van der Put and M. F. Singer. *Galois Theory of Difference Equations, Lecture Notes in Mathematics*, volume 1666. Springer, 1997.
22. M. van der Put and M. F. Singer. *Galois Theory of Linear Differential Equations, Grundlehren der Mathematischen Wissenschaften*, volume 328. Springer, Heidelberg, 2003.
23. M. van Hoeij. Rational solutions of the mixed differential equation and its application to factorization of differential operators. In *Proc. ISSAC'96*, pages 219–225, Zurich, Switzerland, 1996.
24. M. Wu. *On Solutions of Linear Functional Systems and Factorization of Modules over Laurent–Ore Algebras*. PhD thesis, Chinese Academy of Sciences and Université de Nice, (2005).
25. S. Zampieri. A solution of the Cauchy problem for multidimensional discrete linear shift–invariant systems. *Linear Algebra and Its Applications*, 202:143–162, 1994.
26. M. Zhou and F. Winkler. Gröbner bases in difference–differential modules. In *Proc. ISSAC'2006*, pages 353–360, Genova, Italy, 2006.

THE VECTOR RATIONAL FUNCTION
RECONSTRUCTION PROBLEM

ZACH OLESH and ARNE STORJOHANN

David R. Cheriton School of Computer Science
University of Waterloo, Ontario, Canada N2L 3G1
E-mail: astorjoh@uwaterloo.ca
http://www.cs.uwaterloo.ca/~astorjoh/

The final step of some algebraic algorithms is to reconstruct the common denominator d of a collection of rational functions v_*/d from their polynomial images modulo m. Using elementwise rational reconstruction requires that $\deg m > N + D$, where N and D are such that $\deg v_* \leq N$ and $\deg d \leq D$. We present an algorithm, based on minimal approximant basis computation, that can perform the reconstruction for many problem instances even when the modulus has considerably smaller degree, for example $\deg m > N + D/k$ for k a small constant.

Keywords: Rational function reconstruction

1. Introduction

Many algorithms in computer algebra that compute with polynomials from $K[x]$, K a field, use a homomorphic imaging scheme to avoid intermediate expression swell, to allow for simple course-grained parallelization, or to incorporate an output sensitive approach. Often, the last step of these algorithms is to reconstruct the common denominator d of a collection of rational functions $(v_i/d)_{1 \leq i \leq n}$ from their polynomial images $(u_i)_{1 \leq i \leq n}$ modulo m. The images modulo m are typically computed by combining multiple smaller images using either Chinese remaindering $(m = p_1 p_2 \cdots p_l)$ or p-adic lifting $(m = p^l)$.

Typically, the overall cost of an algorithm that uses homomorphic imaging depends on l, the number of images computed, which is directly related to $\deg m$. Ideally, the algorithm computes just enough images to allow reconstruction of the common denominator d. We first recall how elementwise rational function reconstruction can be applied, and then discuss our vector based variant that for some applications can save close to half of the required image computations.

The rational function reconstruction problem takes as input a nonzero modulus $m \in K[x]$, a single image polynomial $u \in K[x]$ with $\deg u < \deg m$, and degree bounds $0 \leq N < \deg m$ and $0 \leq D < \deg m$. A solution to the problem is a pair of polynomials (d, v) such that

$$du \equiv v \bmod m, \quad \deg d \leq D, \quad \deg u \leq N. \tag{1}$$

If (d, v) is a solution to (1) that satisifies $\gcd(d, m) = 1$, then

$$u \equiv \frac{v}{d} \bmod m, \quad \deg d \leq D, \quad \deg u \leq N. \tag{2}$$

For convenience, in order to avoid some special cases, we have used the weaker condition (1) to define a solution to the problem rather than (2).

The vector generalization of the problem is defined similarly except with u replaced by $[u_1, \ldots, u_n] \in K[x]^{1 \times n}$. A solution to the vector version is then a pair $(d, [v_1, \ldots, v_n])$ such that

$$d[u_1, u_2, \ldots, u_n] \equiv [v_1, v_2, \ldots, v_n] \bmod m, \quad \deg d \leq D, \quad \deg v_* \leq N. \tag{3}$$

Similarly, if $\gcd(d, m) = 1$ we have

$$[u_1, u_2, \ldots, u_n] \equiv [\frac{v_1}{d}, \frac{v_2}{d}, \ldots, \frac{v_n}{d}] \bmod m, \quad \deg d \leq D, \quad \deg v_* \leq N. \tag{4}$$

The link between solutions of (1) and certain rows of the traditional extended Euclidean algorithm has been well studied.[18] In general, we require $\deg m > N + D$ to ensure that the solution space is uniquely generated, that is, that every solution can be expressed as a polynomial multiple of a single generating solution (d, v). Let $\mathrm{Ratrecon}(u, m, N, D)$ denote a function that takes as input an instance of the problem with $\deg m > N + D$, and returns as output the first component d (possibly the zero polynomial) of a generating solution. The approach taken in various software libraries[6,8,10,17] to compute the common d of the vector version of the problem is as follows:

Choose $N \geq 0$ and $D \geq 0$ such that $\deg m > N + D$;
$d := 1$;
for i **from** 1 **to** n **do**
    $d := d \times \mathrm{Ratrecon}(du_i \bmod m, m, N, D)$
od;
return d

The choice of N and D will depend on the particular application. Suppose that the v_i and d shown in (4) are the actual target solution to a particular problem. On the one hand, if N and D are *a priori* bounds satisfying (4), then we know *a priori* that the output of the code fragment will be the same denominator d (up to normalization). On the other hand, if an output sensitive approach is being used, and N and D are guesses which may or may not satisfy (4), then the output must be assayed for correctness. If determined to be incorrect, the modulus m is augmented and the reconstruction attempted again. Implementations of Ratrecon, using either the algorithm of this paper or an approach based on half-gcd,[18] have running time bounded by $O(\mathsf{B}(\deg m))$ operations in K, where B is a cost function for gcd-like operations[a]. Thus, the code fragment above will solve the vector version of the problem with $O(n\,\mathsf{B}(\deg m))$ operations in K. Note that the running time for the reconstruction is pseudo-linear in the size of the input; in typical applications the cost of computing the images $[u_1, \ldots, u_m] \bmod m$ will dominate, even to the extent that the time for the reconstruction is negligible in comparison. To save on the number of images that are computed and thus speed up the overall computation we must relax the condition $\deg m > N + D$.

Suppose $\deg m > N + D/k$ for some $k \in \mathbb{Z}_{>0}$. We present an algorithm that computes a complete basis of solutions to (3) using

$$O(nk^{\omega-1}\,\mathsf{B}(\deg m)) \tag{5}$$

operations in K, where $2 \leq \omega \leq 3$ is a feasible exponent for matrix multiplication. By a basis we mean a set of solutions $(d^{(i)}, v^{(i)})_{1 \leq i \leq s}$, each $d^{(i)} \in \mathsf{K}[x]$ and $v^{(i)} \in \mathsf{K}[x]^{1 \times n}$, such that every solution admits a unique decomposition as a $\mathsf{K}[x]$-linear combination of basis elements. The algorithm is similar to the approach based on Ratrecon above, except with the loop iterating only n/k times, each iteration dealing with a block of k images simultaneously. The approach works because we can show that the solution basis for all subproblems will have dimension bounded by k. Actually, for many problem instances the solution space will be uniquely generated ($s \leq 1$) whenever $\deg m > N + D/n$. Next we give an example of an application that generates such problem instances.

Suppose we want to compute $A^{-1}b \in \mathsf{K}(x)^{n \times 1}$ for a nonsingular $A \in \mathsf{K}[x]^{n \times n}$ and $b \in \mathsf{K}[x]^{n \times 1}$ from the image $A^{-1}b \bmod m$ for some m. For simplicity, assume that $\deg b = \deg A$. Let N be a bound for the degree of the numerators of $A^{-1}b$. For example, the *a priori* bound $N = n \deg A$ will be tight for a generic problem instance. From the assumption that $\deg b =$

---

[a]We can take $\mathsf{B}(t) = \mathsf{M}(t) \log t$ where M is a multiplication time for $\mathsf{K}[x]$, see [18, Def. 8.26].

deg A it follows that the denominator of $A^{-1}b$ also has degree bounded by N. To apply elementwise reconstruction we need the image $A^{-1}b \bmod m$ with $\deg m > 2N$. However, in Ref. 16 it was observed that output of the vector rational reconstruction problem with input $A^{-1}b \bmod m$ will be uniquely generated whenever $\deg m > N + \deg A$. Thus, using the vector reconstruction algorithm it will suffice to have $\deg m > N + N/k$ for any $k \in \mathbb{Z}_{>0}$ that satisfies $N/k \geq \deg A$. For k a small constant, say $k = 5$, the reconstruction will still be relatively fast (compare with (5)) but the required lower bound $N + N/5$ for the modulus degree is a factor of 0.6 smaller than the bound $2N$ required for the elementwise approach.

We defined the rational function and vector rational function reconstruction problem to take as input bounds N and D. We remark that algorithms for a variant of the first problem called maximal quotient rational function reconstruction are given in Refs. 12,15. The maximal quotient problem takes as input u and m but not N and D, and returns as output the most likely candidate for v/d. The maximal quotient algorithms are useful in conjunction with an output sensitive approach when the difference between $\deg v$ and $\deg d$ may be large, but unknown. In particular, the approach is likely to succeed when $\deg m$ is modestly larger than $\deg v + \deg d$, compared to the required $\deg m > 2 \max(\deg v, \deg d)$ when a common bound $N = D$ is specified.

The rest of this paper is organised as follows. Sections 2 and 3 recall the notion of a reduced basis and minimal approximant bases. Section 3 also gives an algorithm for a special type of simultaneous matrix Padé approximation, the basis of the vector rational function reconstruction algorithm presented in Sec. 4. In Sec. 5 we show how the vector reconstruction algorithm may be applied to rational system solving over $\mathsf{K}[x]$. For more background on the definitions and concepts introduced in Secs. 2 and 3 we refer to Refs. 1–4. Fundamental notions and algorithms for polynomial matrices can be found in Refs. 5,11.

2. Reduced bases

Let $A \in \mathsf{K}[x]^{n \times m}$ have rank r. Let $\mathcal{L}(A)$ denote the lattice generated by the set of all $\mathsf{K}[x]$-linear combinations of rows of A. In many applications we are interested in the subset of a lattice comprised of all rows $w \in \mathsf{K}[x]^{1 \times m}$ that satisfy a degree constraint specified by a fixed multi-index $\vec{n} = (n_1, n_2, \ldots, n_m) \in \mathbb{Z}^m$:

$$w = [\overset{\leq n_1}{w_1}, \overset{\leq n_2}{w_2}, \ldots, \overset{\leq n_m}{w_m}] \in \mathsf{K}[x]^{1 \times m} \tag{6}$$

Following [1, Def. 3.1], the *defect* of a row $w = [w_1, w_2, \ldots, w_m] \in \mathsf{K}[x]^{1 \times m}$ with respect to $\vec{n}$ is defined by

$$\mathrm{dct}(w) = \mathrm{dct}(w, \vec{n}) := \min_i \{n_i + 1 - \deg w_i\}, \tag{7}$$

where the zero polynomial has degree $-\infty$. The notion of defect measures the gap between w and the degree constraint $\vec{n}$: w satisfies (6) if and only if $\mathrm{dct}(w)$ is positive. The following definition is similar to [3, Def. 5.1].

Definition 2.1. A matrix $B = [\, b_1^T \mid b_2^T \mid \ldots \mid b_r^T \,]^T \in \mathsf{K}[x]^{r \times m}$ is a *reduced basis* of type $\vec{n}$ for $A \in \mathsf{K}[x]^{n \times m}$ if the following conditions are satisfied:

(i) B has full row rank and $\mathcal{L}(B) = \mathcal{L}(A)$. [**basis property**]
(ii) Each $w \in \mathcal{L}(B)$ admits a unique decomposition $w = \sum_{i=1}^r c_i b_i$ with $c_i \in \mathsf{K}[x]$, $\deg c_i \le \mathrm{dct}(b_i) - \mathrm{dct}(w)$, $1 \le i \le r$. [**reduced property**]

The reduced bases are precisely those with maximal defect.
By *positive part* of a reduced basis we mean the submatrix comprised of the rows with positive defect. All $w \in \mathcal{L}(A)$ that satisfy the degree constraint $\vec{n}$ are generated by the positive part of a reduced basis for A: if $\mathrm{dct}(b_i) \le 0$ and $\mathrm{dct}(w) > 0$, then the c_i of Def. 2.1 has $\deg c_i \le \mathrm{dct}(b_i) - \mathrm{dct}(w) < 0$ and thus c_i is the zero polynomial.
Suppose B is a basis for A, rows permuted so that defects are nonincreasing. Then reduced bases are precisely those with $(\mathrm{dct}(b_1), \ldots, \mathrm{dct}(b_r))$ lexicographically maximal among all bases for A whose rows are similarly permuted. Thus, up to row permutation, any two reduced bases of type $\vec{n}$ for A will have the same tuple of defects. It follows that the number of rows in the positive part of a reduced basis is an invariant of A.

3. Minimal approximant bases

Let $G \in \mathsf{K}[x]^{n \times m}$, $\vec{n} \in \mathbb{Z}^n$, and $d \in \mathbb{Z}_{\ge 0}$.

Definition 3.1. An order d *minimal approximant* of type $\vec{n}$ for G is a reduced basis M of type $\vec{n}$ for the lattice $\{w \in \mathsf{K}[x]^{1 \times n} \mid wG \equiv 0 \bmod x^d\}$.

Note that a minimal approximant M as in Def. 3.1 will necessarily have dimension $n \times n$, be nonsingular, and satisfy $MG \equiv 0 \bmod x^d$.
The following is restatement of [9, Theorem 2.4]. We remark that Ref. 9 gives more precise cost estimates in terms of certain ad hoc cost functions. We will use the exponent ω and cost function B.

Theorem 3.1. *There exists an algorithm* MinBasis *that takes as input* $(G, d, \vec{n}) \in (\mathsf{K}[x]^{n \times m}, \mathbb{Z}_{\ge 0}, \mathbb{Z}^n)$ *and returns as output* $(M, \delta) \in$

$(\mathsf{K}[x]^{n \times n}, \mathbb{Z}^n)$, an order d minimal approximant M of type $\vec{n}$ for G together with a tuple $\delta = (\delta_1, \ldots, \delta_n)$ of the defects of rows of M. If $m \leq n$, the cost of the algorithm is $O(n^\omega \, \mathsf{B}(d))$ operations in K.

For brevity, we will say that (M, δ) in Theorem 3.1 solves the minimal approximant problem with input $(G, d, \vec{n})$. By $\texttt{PosMinBasis}(G, d, \vec{n})$ we mean the output of $\texttt{MinBasis}(G, d, \vec{n})$ restricted to the rows with positive defect; this may be a $0 \times n$ matrix.

We now give two technical lemmas that follow from the definition of minimal approximant and the properties of reduced bases. The first lemma states that zero rows in an input matrix can be ignored as far as minimal approximant basis computation is concerned.

$$\overset{H}{\left[\begin{array}{c|c} M & \\ \hline & I_k \end{array} \right] \left[\begin{array}{c} * \\ \hline - \end{array} \right] \equiv 0 \bmod x^d}$$

Lemma 3.1. Let $H \in \mathsf{K}[x]^{n \times m}$ have its last k rows zero and let $\vec{n} = (n_1, \ldots, n_n)$. If $M \in \mathsf{K}[x]^{(n-k) \times (n-k)}$ is an order d minimal approximant of type $(n_1, \ldots, n_{n-k})$ for the first $n - k$ row of H, then $\mathrm{diag}(M, I_k)$ is an order d the minimal approximant of type $\vec{n}$ for H.

The next lemma follows as a special case of [2, Theorem 5.1], which gives a general result regarding the recursive computation of minimal approximants. Let $\mathbf{1}$ denote the tuple $(1, 1, \ldots, 1)$ of appropriate length.

Lemma 3.2. Let $H \in \mathsf{K}[x]^{n \times m}$ and $H' \in \mathsf{K}[x]^{n \times m'}$. If $(M, \delta) := \texttt{MinBasis}(H, d, \vec{n})$ and $(M', \delta') := \texttt{MinBasis}(MH', d, \delta - 1)$, then $(M'M, \delta')$ solves the minimal approximant problem with input $([H|H'], d, \vec{n})$.

The -1 in the second call to $\texttt{MinBasis}$ in Lemma 3.2 is due to the $+1$ in the definition of defect (see (7)). For example, in the special case where H is the zero matrix, an order d minimal approximant of type $\vec{n}$ for H is given by I_n, with row defects $\delta = \vec{n} + \mathbf{1}$. For more details we refer to [2, Sections 3 and 4].

As noted after Def. 2.1, if $w \in \mathcal{L}(\texttt{MinBasis}(H, d, \vec{n}))$ has positive defect with respect to $\vec{n}$, then $w \in \mathcal{L}(\texttt{PosMinBasis}(H, d, \vec{n}))$. Since $\mathcal{L}(M'M) \subseteq \mathcal{L}(M)$, any row in $M'M$ with positive defect with respect to $\vec{n}$ is comprised of a linear combination of rows of $\texttt{PosMinBasis}(H, d, \vec{n})$. We get the following as a corollary.

Corollary 3.1. Lemma 3.2 still holds if *MinBasis* is replaced by *PosMinBasis* and "minimal approximant" is replaced by "positive part minimal approximant."

3.1. *An algorithm for simultaneous Padé approximation*

We describe an algorithm to compute an order d minimal approximant of type $\vec{n}$ for an input matrix G that can be decomposed as

$$G = \begin{bmatrix} \begin{array}{c|c|c|c} G_1 & G_2 & \cdots & G_n \\ \hline E & & & \\ \hline & E & & \\ \hline & & \ddots & \\ \hline & & & E \end{array} \end{bmatrix} \in \mathsf{K}[x]^{(m+tn)\times nk}, \tag{8}$$

each $G_i \in \mathsf{K}[x]^{m\times k}$ and $E \in \mathsf{K}[x]^{t\times k}$. We will assume that $\vec{n} = (\vec{n}_1, \vec{n}_2, .^{\,n}\!.$ $, \vec{n}_2)$ with $\vec{n}_1 \in \mathbb{Z}_{\geq 0}^m$ and $\vec{n}_2 \in \mathbb{Z}_{\geq 0}^t$, but remark that the algorithm we present can be adapted to work for an arbitrary degree constraint $\vec{n} \in \mathbb{Z}^{m+tn}$. Actually, our goal is to compute only the first m columns of the positive part of an order d minimal approximant of type $\vec{n}$. Lemma 3.1 and Corollary 3.1 suggest an iterative approach that works in stages for $i = 1, 2, \ldots, n$. The approach can be understood by considering stage 2. Suppose we have the first m columns $\bar{M} \in \mathsf{K}[x]^{s\times m}$ of the positive part $[\,\bar{M} \mid *\,] \in \mathsf{K}[x]^{s\times(m+t)}$ of an order d minimal approximant of type $(\vec{n}_1, \vec{n}_2)$ for

$$\begin{bmatrix} G_1 \\ E \end{bmatrix} \in \mathsf{K}[x]^{(m+t)\times k},$$

together with a corresponding tuple $\delta \in \mathbb{Z}_{>0}^s$ of defects. By Lemma 3.1, $\mathrm{diag}([\,\bar{M} \mid *\,], I_t)$, with defect tuple $(\delta, \vec{n}_2 + 1)$, is the the positive part of an order d minimal approximant of type $(\vec{n}_1, \vec{n}_2, \vec{n}_2)$ for the first k columns H of

$$[\,H \mid H'\,] = \begin{bmatrix} \begin{array}{c|c} G_1 & C_2 \\ \hline E & \\ \hline & E \end{array} \end{bmatrix} \in \mathsf{K}[x]^{(m+2t)\times 2k}. \tag{9}$$

By Corollary 3.1, if

$$(M', \delta') := \mathtt{PosMinBasis}(\mathrm{diag}([\,\bar{M} \mid *\,], I_t)H', d, (\delta, \vec{n}_2 + 1) - 1),$$

then $M'\mathrm{diag}([\,\bar{M} \mid *\,], I_t)$ will be the positive part of an order d minimal approximant of type $(\vec{n}_1, \vec{n}_2, \vec{n}_2)$ for $[\,H \mid H'\,]$. The key observation is that the first argument of $\mathtt{PosMinBasis}$ is given by

$$\begin{bmatrix} \begin{array}{c|c} \bar{M} & * \\ \hline & I_t \end{array} \end{bmatrix} \begin{bmatrix} G_2 \\ \hline E \end{bmatrix} = \begin{bmatrix} \bar{M}G_2 \\ \hline E \end{bmatrix},$$

so we don't need to know the unknown block $*$ of $[\bar{M} \mid *]$. Once M' is computed, the first m columns of the positive part of a minimal approximant for $[H \mid H']$ can be computed as $M'\bar{M}$. Stages $i = 3, 4, \ldots, n$ are similar. This gives the following algorithm.

Algorithm: $\text{SimPade}([G_1, \ldots, G_n], E, d, \vec{n}_1, \vec{n}_2)$
Input: $G_* \in \mathsf{K}[x]^{m \times k}$, $E \in \mathsf{K}[x]^{t \times k}$, $d \in \mathbb{Z}_{\geq 0}$, $\vec{n}_1 \in \mathbb{Z}_{\geq 0}^m$, $\vec{n}_2 \in \mathbb{Z}_{\geq 0}^t$.
Output: $(\bar{M}, \delta)$, $\bar{M}$ the first m columns of an M such that (M, δ) is a valid output of $\text{PosMinBasis}(G, d, (\vec{n}_1, \vec{n}_2, \overset{n}{\ldots}, \vec{n}_2))$, with G as in (8).

$(\bar{M}, \delta) := (I_m, \vec{n}_1 + 1)$;
for i **from** 1 **to** n **do**
$\quad \delta := (\delta, \vec{n}_2 + 1)$;
$\quad (M', \mu) := \text{PosMinBasis}\left(\left[\dfrac{\bar{M}G_i}{E}\right], d, \delta - 1\right)$;
$\quad \bar{M} := M'\bar{M}$
od;
return $(\bar{M}, \delta)$

The cost of algorithm SimPade will depend on the row dimensions of the first argument to the n calls to PosMinBasis. In the next section we will see that for some inputs to the algorithm we can be sure that $\bar{M}$ will never have more than k rows.

Theorem 3.2. *Algorithm* SimPade *is correct. If* $t = O(k)$ *and the dimension of* $\bar{M}$ *remains bounded by* k *throughout, the cost of the algorithm is* $O((nk + m)k^{\omega-1}\,\mathsf{B}(d))$ *operations in* K.

4. Vector rational function reconstruction

Fix the following quantities throughout this section:

- a nonzero modulus $m \in \mathsf{K}[x]$,
- an input vector $u \in \mathsf{K}[x]^{1 \times n}$ with $\deg u < \deg m$, and
- degree bounds N and D with $0 \leq N < \deg m$ and $0 \leq D < \deg m$.

A vector $[d \mid v] \in \mathsf{K}[x]^{1 \times (n+1)}$ ($d \in \mathsf{K}[x]$, $v \in \mathsf{K}[x]^{1 \times n}$) solves the vector rational function reconstruction problem if $du \equiv v \bmod m$, with $\deg d \leq D$ and $\deg v \leq N$. The complete set of solutions is thus

$$\mathcal{S} = \{[d \mid v] \in \mathsf{K}[x]^{1 \times (n+1)} \mid du \equiv v \bmod m, \ \deg d \leq D, \ \deg v \leq N\}.$$

Consider the lattice generated by the nonsingular matrix

$$A = \left[\begin{array}{c|c} 1 & u \\ \hline & mI_n \end{array} \right] \in \mathsf{K}[x]^{(n+1)\times(n+1)}. \tag{10}$$

Any vector in $\mathcal{L}(A)$ with degree strictly less than $\deg m$ has the form $[\, d \,|\, du \bmod m \,]$ for $d \in \mathsf{K}[x]$ with $\deg d < \deg m$: the rows of A containing mI_n serve to reduce modulo m the last n entries in $d\,[\,1\,|\,u\,]$. If we set degree constraints $(D, N, \ldots, N)$, then $[\, d \,|\, v\,] \in \mathcal{S}$ if and only if $[\, d \,|\, v \,] \in \mathcal{L}(A)$ with $\mathrm{dct}([\, d \,|\, v\,]) > 0$. Thus, $\mathcal{S}$ is generated by the positive part $B = [\, b_1^T \,|\, b_2^T \,|\, \ldots \,|\, b_s^T \,]^T \in \mathsf{K}[x]^{s\times(n+1)}$ of a reduced basis of type $(D, N, \ldots, N)$ for A.

Theorem 4.1. $\mathcal{S} = \{\sum_{i=1}^{s} c_i b_i \mid c_i \in \mathsf{K}[x], \ \deg c_i < \mathrm{dct}(b_i), \ 1 \le i \le s\}$.

Corollary 4.1. *If $e \in \mathsf{K}[x]^{s\times 1}$ is the first column of the positive part of a reduced basis of type $(D, N, \ldots, N)$ for A, then $[\, e \,|\, eu \bmod m \,] \in \mathsf{K}[x]^{s\times n}$ is the positive part of a reduced basis of type $(D, N, \ldots, N)$ for A.*

The next theorem gives an *a priori* upper bound on s, the number of rows in the positive part of a reduced basis of type $(D, N, \ldots, N)$ for A. Since the bound does not depend on n, it also applies for the number of rows in the positive part of a reduced basis of type $(D, N, \ldots, N)$ for the leading $j \times j$ submatrix of A, for any j with $2 \le j \le n+1$.

Theorem 4.2. $s \le k$ *for $k \in \mathbb{Z}_{>0}$ minimal such that $\deg m > N + D/k$.*

Proof. Assume for now that $N \ge D$. Then R is a reduced basis of type $(D, N, \ldots, N)$ for A if and only if $R' := R \operatorname{diag}(x^{N-D}, I_n)$ is a reduced basis of type $(N, N, \ldots, N)$ for $A' := A \operatorname{diag}(x^{N-D}, I_n)$. Thus, s is equal to the number of rows in R' with degree at most N. A reduced basis of type $(N, N, \ldots, N)$ for A' will have degree at most $\deg A'$, so $\deg R' \le \deg A' = \deg m$. Using the fact that the determinant of a polynomial matrix is bounded by the sum of the row degrees now gives

$$\deg \det R' \le sN + (n+1-s) \deg m. \tag{11}$$

Using the fact that $\det R'$ is a scalar multiple of $\det A'$ gives

$$\deg \det R' = \deg \det A' = N - D + n \deg m. \tag{12}$$

Combining (11) and (12) and solving for $\deg m$ gives

$$\deg m \le N + \frac{D}{s-1}.$$

It follows that $s - 1 < k$. The case $D > N$ is similar. $\square$

Let

$$G = \begin{bmatrix} u \\ \hline -I_n \\ \hline mI_n \end{bmatrix} \in \mathsf{K}[x]^{(2n+1)\times n}.$$

Dependant on the assumption that $\deg u < \deg m$, each $[\,d\,|\,v\,] \in \mathcal{S}$ can be extended with $r := -(du - v)/m \in \mathsf{K}[x]^{1\times n}$ such that $\deg r \leq D - 1$ and $[\,d\,|\,v\,|\,r\,]G = 0$. Conversely, if $[\,d\,|\,v\,|\,r\,] \in \mathsf{K}[x]^{1\times(2n+1)}$ satisfies $[\,d\,|\,v\,|\,r\,]G \equiv 0 \bmod x^{D+\deg m}$ and $(\deg d, \deg v, \deg r) \leq (D, N, D-1)$, then $du - v + mr \equiv 0 \bmod x^{D+\deg m}$ with $\deg(du - v + mr) < D + \deg m$, implying $du - v + mr = 0$ and thus $[\,d\,|\,v\,] \in \mathcal{S}$. Thus, the first $n + 1$ columns of the positive part of an order $D + \deg m$ minimal approximant of type

$$(D, N, .^n_{.}., N, D - 1, .^n_{.}., D - 1) \tag{13}$$

for G is a reduced basis of type $(D, N, \ldots, N)$ for A. By Corollary 4.1, it will suffice to compute only the first column of such a minimal approximant.

To apply algorithm $\texttt{SimPade}$ we need to adjust the matrix G slightly. Let k be either n or as in Theorem 4.2, whichever is minimal. Assume for now that k divides n and write $u = [\,u_1\,|\,u_2\,|\,\ldots\,|\,u_{n/k}\,]$, each $u_* \in \mathsf{K}[x]^{1\times k}$. Permute the last $2n$ rows of G so that

$$G = \left[\begin{array}{c|c|c|c} u_1 & u_2 & \cdots & u_{n/k} \\ \hline -I_k & & & \\ mI_k & & & \\ \hline & -I_k & & \\ & mI_k & & \\ \hline & & \ddots & \\ \hline & & & -I_k \\ & & & mI_k \end{array}\right] \in \mathsf{K}[x]^{(2n+1)\times n}.$$

In the special case when m is a power of x, the vector rational reconstruction problem is a simultaneous Padé approximation problem: the positive part of a reduced basis for A shown in (10) is the positive part of an order $\deg m$ minimal approximant of type $(D, N, \ldots, N)$ for

$$G = \left[\begin{array}{c|c|c|c} u_1 & u_2 & \cdots & u_{n/k} \\ \hline -I_k & & & \\ \hline & -I_k & & \\ \hline & & \ddots & \\ \hline & & & -I_k \end{array}\right] \in \mathsf{K}[x]^{(n+1)\times n}.$$

This shows correctness of the following algorithm.

Algorithm: VectorRecon(u, m, N, D)
Input: $u \in K[x]^{1 \times n}$, nonzero $m \in K[x]$, $N \in \mathbb{Z}_{\geq 0}$, $D \in \mathbb{Z}_{\geq 0}$.
Output: An $e \in K[x]^{s \times 1}$ as in Corollary 4.1.
Condition: $N < \deg m$, $D < \deg m$, $\deg u < \deg m$.

$k := \min\{n, \min\{t \in \mathbb{Z}_{>0} \mid \deg m > N + D/t\}\}$;
Augment u with at most $k - 1$ zeros so that $k \mid n$;
if $m = x^{\deg m}$ **then**
$\quad E := -I_k$;
$\quad \vec{n}_2 := (N, .\overset{k}{.}., N)$;
$\quad d := \deg m$
else
$$E := \begin{bmatrix} -I_k \\ mI_k \end{bmatrix};$$
$\quad \vec{n}_2 := (N, .\overset{k}{.}., N, D - 1, .\overset{k}{.}., D - 1)$;
$\quad d := D + \deg m$
fi;
$\vec{n}_1 := (D)$;
Write $u = [\, u_1 \mid u_2 \mid \ldots \mid u_{n/k} \,]$, each $u_i \in K[x]^{1 \times k}$;
$(e, *) := \mathtt{SimPade}([u_1, u_2, \ldots, u_n]), E, d, \vec{n}_1, \vec{n}_2)$;
Normalize each entry in e to be monic;
return e

Theorem 4.3. *Algorithm* VectorRecon *is correct. The cost of the algorithm is* $O(nk^{\omega-1} \mathsf{B}(\deg m))$ *operations in* K, *where* $k \in \mathbb{Z}_{>0}$ *is minimal such that* $\deg m > N + D/k$.

5. Application to linear solving

Let a nonsingular $A \in K[x]^{n \times n}$ and $b \in K[x]^{n \times 1}$ be given. Let $d \in K[x]$ be the denominator of $A^{-1}b$, that is, the minimal degree monic polynomial such that $v := dA^{-1}b$ is over $K[x]$. One of the most effective methods to compute d is to iteratively compute

$$u := A^{-1}b \bmod p^l = c_0 + c_1 p + c_2 p^2 + \cdots + c_{l-1}p^{l-1}, \qquad (14)$$

each $c_i \in K[x]^{n \times 1}$ with $\deg c_i < \deg p$, for larger and larger l using p-adic lifting[7,14] for some p with $\gcd(p, \det A) = 1$, and then apply rational reconstruction. If desired, v can be computed as $du \bmod m$ once d is found. In the following theorem m plays the role of p^l.

Theorem 5.1. *If* $\deg m > \max(N+\deg A, D+\deg b)$ *and* $u = A^{-1}b \bmod m$
then the output of VectorRecon(u^T, m, N, D) *is either:*

(i) $e = [\,d\,] \in K[x]^{1\times 1}$, *if* $N \geq \deg v$ *and* $D \geq \deg d$, *or*
(ii) $e \in K[x]^{0\times 1}$, *if at least one of* $N < \deg v$ *or* $D < \deg d$.

Proof. Suppose $e = [e_1, e_2, \ldots, e_s]^T \in K[x]^{s\times 1}$ is the output of
VectorRecon, and for $1 \leq i \leq s$ let $v_i := e_i A^{-1}b \bmod m$. The s vectors $[\,e_i \mid v_i^T\,] \in K[x]^{1\times(n+1)}$ are linearly independent and satisfy $Av_i \equiv e_i b \bmod m$.
Since $\max(\deg Av_i, \deg e_i b) \leq \max(N + \deg A, D + \deg b) < \deg m$, we ac-
tually have $Av_i = e_i b$. Parts (i) and (ii) now follow by noting that the
dimension of the solution space for these cases are 1 and 0, respectively. $\square$

Suppose N and D are *a priori* bounds: $N \geq \deg v$ and $D \geq \deg d$. Standard
rational function reconstruction[18] can be used to recover d in $O(n\,\mathsf{B}(\deg m))$
field operations but requires $\deg m > N + D$. By Theorem 5.1, Algo-
rithm VectorRecon can recover d in $O(nk^{\omega-1}\mathsf{B}(\deg m))$ field operations
where $\deg m > \max(N + \deg A, D + \deg b, N + D/k)$.
Algorithm VectorRecon can also be used in conjuction with an output
sensitive approach. Let $m = p^l$ and suppose we have u as in (14). Set $\bar{N}$
to be the maximal integer such that $\deg m > \bar{N} + \max(\deg A, \deg b, \bar{N}/k)$.
According to Theorem 5.1, the call VectorRecon($u^T, m, \bar{N}, \bar{N}$) will either
recover the denominator d or determine that $\max(\deg d, \deg v) > \bar{N}$.

6. Conclusion

The approach we have described here for reconstructing a vector of rational functions with common denominator can be adapted to the problem
of reconstructing a vector of rational numbers with a common denominator. This requires the use of integer lattice basis reduction[13] and will be
described in a future paper.

References

1. B. Beckermann and G. Labahn, *SIAM Journal on Matrix Analysis and Applications* **15**, 804 (1994).
2. B. Beckermann and G. Labahn, *Journal of Computational and Applied Math*
77, 5 (1997).
3. B. Beckermann, G. Labahn and G. Villard, Shifted normal forms of polynomial matrices, in *Proc. Int'l. Symp. on Symbolic and Algebraic Computation:
ISSAC '99*, ed. S. Dooley (ACM Press, New York, 1999).

4. B. Beckermann, G. Labahn and G. Villard, *Normal Forms for General Polynomial Matrices*, Research Report 2002–1, ENS Lyon (France, 2002).

5. D. Bini and V. Y. Pan, *Polynomial and Matrix Computations, Vol 1: Fundamental Algorithms* (Birkhauser, Boston, 1994).

6. Z. Chen and A. Storjohann, A BLAS based C library for exact linear algebra on integer matrices, in *Proc. Int'l. Symp. on Symbolic and Algebraic Computation: ISSAC '05*, ed. M. Kauers (ACM Press, New York, 2005).

7. J. D. Dixon, *Numer. Math.* **40**, 137 (1982).

8. J.-G. Dumas, T. Gautier, M. Giesbrecht, P. Giorgi, B. Hovinen, E. Kaltofen, B. D. Saunders, W. J. Turner and G. Villard, LinBox: A generic library for exact linear algebra., in *Proc. First Internat. Congress Math. Software ICMS 2002, Beijing, China*, eds. A. J. Cohen and N. Gao, X.-S. andl Takayama (World Scientific, Singapore, 2002).

9. P. Giorgi, C.-P. Jeannerod and G. Villard, On the complexity of polynomial matrix computations, in *Proc. Int'l. Symp. on Symbolic and Algebraic Computation: ISSAC '03*, ed. R. Sendra (ACM Press, New York, 2003).

10. P. Giorgi, Arithmetic and algorithmic in exact linear algebra for the LinBox library, PhD thesis, Ecole normale superieure de Lyon, LIP, (Lyon, France, 2004).

11. T. Kailath, *Linear Systems* (Prentice Hall, Englewood Cliffs, N.J., 1980).

12. S. Khodadad and M. Monagan, Fast rational function reconstruction, in *Proc. Int'l. Symp. on Symbolic and Algebraic Computation: ISSAC '05*, ed. J.-G. Dumas (ACM Press, New York, 2006).

13. A. K. Lenstra, H. W. Lenstra and L. Lovász, *Math. Ann.* **261**, 515 (1982).

14. R. T. Moenck and J. H. Carter, Approximate algorithms to derive exact solutions to systems of linear equations., in *Proc. EUROSAM '79, volume 72 of Lecture Notes in Compute Science*, (Springer-Verlag, Berlin-Heidelberg-New York, 1979).

15. M. Monagan, Maximal quotient rational reconstruction: an almost optimal algorithm for rational reconstruction, in *Proc. Int'l. Symp. on Symbolic and Algebraic Computation: ISSAC '04*, ed. J. Gutierrez (ACM Press, New York, 2004).

16. T. Mulders and A. Storjohann, Rational solutions of singular linear systems, in *Proc. Int'l. Symp. on Symbolic and Algebraic Computation: ISSAC '00*, ed. C. Traverso (ACM Press, New York, 2000).

17. V. Shoup, *NTL: A Library for Doing Number Theory*, (2005). http://www.shoup.net/ntl/.

18. J. von zur Gathen and J. Gerhard, *Modern Computer Algebra*, 2 edn. (Cambridge University Press, 2003).

FAST ALGORITHM FOR COMPUTING MULTIPOLE MATRIX ELEMENTS WITH LEGENDRE POLYNOMIALS

V.YU. PAPSHEV, S.YU. SLAVYANOV

Department of Computational Physics, St-Petersburg State University,
Botanicheskaya 1, 198506, St-Petersburg, Russia,
E-mail: slav@ss2034.spb.edu

The differential equation for the product of Legendre polynomials provides a recursive algorithm for calculation succeeding multipole matrix elements with them. Non-multipole matrix elements as, for instance, particular Clebsh-Gordon coefficients also can be computed on the basis of this algorithm.

Keywords: matrix elements, Legendre polynomials, Clebsh-Gordon coefficients

Introduction

By multipole matrix elements are meant integrals of the form

$$< y|x^k|u > = \int_a^b y_n(x)u_m(x)x^k dx = \int_a^b v_{nm}(x)x^k dx. \qquad (1)$$

Here $y_n(x)$ u_m – are different eigenfunctions of an appropriate singular Sturm-Liouville problem considered on an interval [a,b] where (a and b – are singularities of the differential equations for y_n and u_m). In those cases, when initial equations are given in a not self-adjoint form integrals (1) are substituted for

$$< y|x^k|u > = \int_a^b \omega(x)y_n(x)u_m(x)x^k dx = \int_a^b \omega(x)v_{nm}(x)x^k dx, \qquad (2)$$

where $\omega(x)$ - is the weight function which helps to transform the equation into the self-adjoint form and appears in the orthogonality condition for the corresponding eigenfunctions.

Any other matrix element of the form

$$< y|f|u > = \int_a^b y_n(x)u_m(x)f(x)dx = \int_a^b v_{nm}(x)f(x)dx. \qquad (3)$$

can be precisely calculated if an algorithm for multipole matrix elements
(1) for successive k is known and the function $f(x)$ can be substituted for
truncated Taylor series valid on the interval $[a, b]$. The accurate estimate
for the error can also be obtained.

The knowledge of the differential equation for the function $v(x)$ (further
called the equation for the product) enables to fasten calculation of integrals
(1), (2). The scheme of calculation is as following: firstly, the equation for
the product is subjected to an integral transform. The selection of this
integral transform depends on the character of the singular points a the b.
If these points are Fuchsian (regular) a certain modification of the Hilbert
transform is taken. If the equation includes an irregular singularity a certain
modification of the Laplace transform is taken. The cases referred to Laplace
transform have been studied in other publications of the authors.[3,5] In this
paper the case of Hilbert transform is studied.

The transformed differential equation is also characterized by singularities
and as it is shown below the coefficients of Taylor series of an appropriate
solution of this equation near an appropriate (Fuchsian) singularity appear
to be the defined above multipole matrix elements. These coefficients for
different integer values of k are connected one with another by recursions.
In the case of two-term recursions explicit expressions for multipole matrix
elements are obtained beginning with the first which could be the normalization integral or the dipole matrix element. In the case of three-term
recursions, four-term recursions ets. the arbitrary multipole matrix element
is expressed in terms of several basic multipole matrix elements.

Sometimes the differential equation for the product includes a parameter
which can be considered as small (large). Asymptotics for matrix elements
can be obtained approximately expanding the recursions in degrees of this
parameter (cf. Ref. 5).

There are several other ways to find the discussed recursions, for instance,
based on generalized Wronskian[1] or algebra of commutators.[6] However the
authors hope that their method is the most general and straightforward.

1. Hilbert transform for solutions of the equation for the product

We introduce an integral transform, which is a particular case of Hilbert
transform. Let $f(z)$ be a function defined on an interval $[-1;1]$ with the
following properties

$$\mathrm{supp} f(z) = [-1, 1], \tag{4}$$

$$f(z) \in C^\infty[-1, 1], \tag{5}$$

$$\int_{-1}^{1} f(z)dz = 0. \tag{6}$$

The last condition is related to orthogonality condition for eigenfunctions. The space of such functions is denoted as S. Clearly that S is a linear space. On the functions f out of S the following transform is determined

$$g(\zeta) = H[f(z)]: \quad g(\zeta) = \int_{-1}^{1} \frac{f(z)}{z - \zeta}, \quad |\zeta| > 1. \tag{7}$$

Functions $g(\zeta)$ are holomorphic functions of the complex variable ζ outside unique circle including infinity. The set of these functions constitute the linear space $\tilde{S}$. In the vicinity of infinity such functions can be expanded in a Taylor series in inverse degrees of ζ

$$g(\zeta) = \sum_{k=0}^{\infty} c_k \zeta^{-l-k-1} \tag{8}$$

where l is a nonnegative integer which value is chosen from the condition

$$\int_{-1}^{1} z^j f(z)dz = 0, \ j = 0, 1, \ldots l - 1, \quad \int_{-1}^{1} z^l f(z)dz \neq 0. \tag{9}$$

The following properties of the transform (7) can be easily verified with the help of integration by parts

$$H[f'] = g', \quad H[zf] = \zeta g. \tag{10}$$

They follow from (4-6) and slightly differ from the properties of the conventional Hilbert transform.

Let $f(z)$ – be a product of two eigenfunctions on [-1,1], multiplied by the weight function $\omega(z)$, i.e. $f(z) = y(z)u(z)\omega(z)$, $z \in [-1, 1]$. Assume that $\omega(z)$ is a polynomial. Outside [-1,1] the function $f(z)$ is defined to be zero. Clearly, such function belongs to S.

Formula (7) can be transformed to

$$g(\zeta) = \int_{-\infty}^{\infty} \frac{f(z)}{z - \zeta}dz = -\frac{1}{\zeta} \int_{-\infty}^{\infty} f(z) \sum_{k=0}^{\infty} \left(\frac{z}{\zeta}\right)^{l+k} dz$$

$$= -\sum_{k=0}^{\infty} \frac{1}{\zeta^{l+k+1}} \int_{-\infty}^{\infty} f(z)z^{l+k}dz \tag{11}$$

Thus matrix elements (1) (or (2)) are coefficients of the expansion of the solution of the equation for the product in the vicinity of infinity. However it is important to choose the proper solution out of four lineary independent solutions.

2. Equation for the product of Legendre polynomials and its Hilbert transform

The equation for the product of Legendre polynomials has been obtained in the previous publication of the authors.[4] Let $v_{nm}(x) = P_n(x)P_m(x)$, then the equation for the product would be

$$(r(rv')')'' + \sigma((rv)'' + (rv')') + \Delta^2 v = 0, \tag{12}$$

where

$$r = 1 - x^2, \ \sigma = n(n+1) + m(m+1), \ \Delta^2 = (n-m)^2(n+m+1)^2$$

and n and m are degrees of Legendre polynomials.

After Hilbert transform we obtain the same equation as (12) (but with other independent variable ζ). Its four solutions are characterized by the following behavior at infinity

$$v_1 \sim \zeta^{n+m}, \quad v_2 \sim \zeta^{n-m-1}, \quad v_3 \sim \zeta^{m-n-1}, \quad v_4 \sim \zeta^{-n-m-2}$$

Let $n > m$, then the solutions v_1, v_2 are not decreasing and the solutions v_3, v_4 are decreasing at infinity. All solutions of Eq. (12) have no branching point at infinity and and therefore the point $z = \infty$ is an apparent singularity for Eq. (12). The solutions v_1 and v_2 can not be generated by Hilbert transform, since they are not holomorphic at infinity.

Any Legendre polynomial $P_n(x)$ is orthogonal to any polynomial of the degree less than n

$$\int_{-1}^{1} z^j P_n(z)P_m(z)dz = 0, \ j < n - m.$$

Hence the value of l in (9) is equal to $n - m$, and only v_3 is characterized by the corresponding behavior at infinity and is generated by Hilbert transform.

From the tables of integral transforms[2] it is known that

$$v_3 = AP_m(\zeta)Q_n(\zeta) \tag{13}$$

where A – is the appropriate constant (actually its value is $A = 2$). The solution v_3 can be searched in the form of a series

$$v_3 = \sum_{k=0}^{\infty} \frac{c_k}{\zeta^{n-m+k+1}}. \tag{14}$$

The coefficients c_k can be found according (13) by multiplication of a polynomial by infinite series. However it is not the fastest algorithm. Our goal is to obtain a three-term recursion. Such recursions for coefficients c_k are found by substitution of series (14) into Eq. (12).

$$((l+k+1)(l+k)^2(l+k-1) - 2\sigma(l+k)^2 + \Delta^2)c_k -$$
$$2((l+k)(l+k-1)^3 - \sigma(l+k)(l+k-1))c_{k-2} + \qquad \cdot$$
$$(l+k)(l+k-1)(l+k-2)(l+k-3)c_{k-4} = 0 \qquad (15)$$

The coefficient c_0 should be calculated explicitly, then from (15) the coefficient c_2 is found and further from recursion (15) the succeeding coefficient $c_4, c_6, \ldots$ are calculated.

Since Legendre polynomials are either even or odd with respect to zero the recursion (15) at even $n-m$ associates even polynomials and at odd $n-m$ associates odd polynomials.

Here is an elementary calculation of c_0.

$$c_0 = \int_{-1}^{1} x^{n-m} P_n(x) P_m(x) dx.$$

From orthogonality of Legendre polynomials it follows

$$c_0 = \int_{-1}^{1} x^n P_n(x) \frac{2m!}{2^m(m!)^2} dx$$

where the numerical factor corresponds to the coefficient in front of the highest degree of $P_m(x)$. By use of Rodrigues formula we obtain

$$c_0 = \frac{2m!}{2^m(m!)^2} \frac{1}{2^n n!} \int_{-1}^{1} x^n \frac{d^n}{dx^n}(1-x^2)^n dx =$$

$$\frac{2m!}{2^m(m!)^2} \frac{(-1)^n(n!)^2 2^n}{(2n+1)!}.$$

The difference second order equation (15) belongs to Poincaré-Perron type. Dividing all coefficients by k^4 we get

$$\alpha_k c_k - 2\beta_k c_{k-2} + \gamma_k c_{k-4} = 0 \qquad (16)$$

with

$$\alpha_k = 1 + O(k^{-2}), \ \beta_k = 1 - 3/k + O(k^{-2}), \ \gamma_k = 1 - 6/k + O(k^{-2}).$$

The corresponding to Eq. (16) characteristic equation has doubly degenerated root equal to unity. The correcting terms move both roots by a quantity of order $O(k^{-2})$ in such a way that one solution increases and the second solution does not increase.

Example: Let $m = 0$ and $n = 2$, so that $l = 2$. The first nonzero matrix element is $c_0 = \frac{4}{15}$. Recurrence (15) in this case is written as

$$((k+2)^2((k+2)^2 - 1) - 12(k+2)^2 + 36)c_k -$$
$$2(k+2)(k+1)((k+1)^2 - 6)c_{k-2} +$$
$$k(k-1)(k+1)(k+2)c_{k-4} = 0, \tag{17}$$

and we find from Eq. (17) other matrix elements $c_2, c_4, c_6, \ldots$

$$c_2 = \frac{8}{35}, \ c_4 = \frac{4}{21}, \ c_6 = \frac{16}{99}, \ c_8 = \frac{20}{143}, \ldots \quad c_{20} = \frac{44}{575}, \ldots$$

Calculated in this way c_k coincide with matrix elements computed with MAPLE. The slight decrease of matrix elements in k is determined by the standard normalization of Legendre polynomials. If they are normalized as

$$\int_{-1}^{1} P_n^2 dx = 1,$$

the decrease would be factorial. Although formula (13) seems to be simple, calculations with its help are more complex than the calculations according to recursion (17). It is known that multiplication of two polynomials takes time more than of order $O(n \ln n)$, where n is the degree of the polynomial. The recurrent calculations (because of its band structure) take the time of order $O(k) + O(n)$, where k is the number of matrix element. Calculations with MAPLE show that, for instance, at $n = 20$ and $k = 500$ the recurrent calculations take at least three times less time than other calculations.

3. Calculation of particular cases of Clebsh-Gordon coefficients

Clebsh-Gordon coefficients is a set of numbers characterized by six indexes arising as a result of integration of three spherical functions. They often appear with respect to summation of angular momentum in quantum mechanics. There exist explicit formulae for Clebsh-Gordon coefficients but they have sufficiently complicated structure and include generalized hypergeometric function. Each formula can be applied for a single coefficient but it hardly can be used for calculation of a set of coefficients. Here a particular case of Clebsh-Gordon coefficients is considered when spherical functions are reduced to Legendre polynomials. We introduce the following integrals

$$\int_{-1}^{1} P_n(x) P_m(x) P_k(x) dx = 2 \begin{pmatrix} m & n & k \\ 0 & 0 & 0 \end{pmatrix}^2. \tag{18}$$

Sometimes another notation is used

$$\begin{pmatrix} m & n & k \\ 0 & 0 & 0 \end{pmatrix} = \frac{(-1)^{m-n}}{\sqrt{2k+1}}(mn00|mnk0). \tag{19}$$

In fact these integrals can be considered as a particular case of (3) with f being a particular polynomial.

Hence, the calculation of the Clebsh-Gordon coefficients can be reduced to a finite summation of those matrix elements which were defined above. The largest benefit in time is obtained if we take maximal integer of m, n, k (let it be k) for outer summation. The following formula for Legendre polynomials is valid

$$P_k(x) = \sum_{j=0}^{[k/2]} d_j x^{k-2j} =$$

$$\sum_{j=0}^{[k/2]} \frac{(-1)^j 2^{-2j} (2k-2j-1)!!}{j!(k-2j)!} x^{k-2j}, \tag{20}$$

where $[k/2]$ is the integer part. Hence we obtain for the square of Clebsh-Gordon coefficients an expression in the form of a series

$$\begin{pmatrix} m & n & k \\ 0 & 0 & 0 \end{pmatrix}^2 = \frac{1}{2} \sum_{j=0}^{[k/2]} d_j < P_m(x)|x^{k-2j}|P_k(x) > . \tag{21}$$

The coefficients d_j should be calculated recursively.

Example. Let $n = 8$, $m = 6$. The first nonzero matrix element is $c_0 = \int_{-1}^{1} x^2 P_n(x) P_m(x) dx = 112/3315$. Then the Clebsh-Gordon coefficients can be calculated with the help of (21) based on the known matrix elements $< P_n(x)|x^{k-2j}|P_m(x) >$ and the coefficients d_j.

$$\tilde{CG}(2) = \frac{56}{1105}, \quad \tilde{CG}(4) = \frac{1008}{46189}, \quad \tilde{CG}(6) = \frac{700}{46189},$$

$$\tilde{CG}(8) = \frac{1200}{96577}, \quad \tilde{CG}(10) = \frac{5544}{482885},$$

$$\tilde{CG}(12) = \frac{2288}{185725}, \quad \tilde{CG}(14) = \frac{22022}{1077205}.$$

The proposed method can be generalized to more complicated cases including, for instance, Heun polynomials.

One of the authors (S.Yu.S.) is grateful to thank Professor S.A. Abramov for permanent friendly support during many years and for the invitation to participate in this Proceedings.

References

1. F. Calogero. Studies in Mathematical Physics. *Essays in Honor of Valentine Bargmann*, eds. E. H. Lieb, B. Simon, A.S. Wightman, 107-150, (Princeton University Press, Princeton, New Jersey, 1976), pp.107-150.
2. I. S. Gradshtein and I. M. Ryzhik. *Tables of sums and integrals*, (Moscow, Fizmatgiz, 1965, in Russian).
3. S.Yu. Slavyanov. Multipole moments for Coulomb and oscillator wavefunctions and the Heun equation. *J. Phys. A* **32**, 1773 (1999).
4. S.Yu. Slavyanov and V. Yu. Papshev. Product of the Jacobi polynomials, in *Ulmer Seminare 2003*, (Ulm, Germany, 2003).
5. S.Yu. Slavyanov and V. Yu. Papshev. Explicit and asymptotic study of one-dimensional multipole matrix elements, in *Nonadiabatic transition in quantum system*, pp. 84-93, (Chernogolovka, Russia, 2004).
6. N. F. Truskova. Linear algebra of integrals in the two-centers problem of quantum mechanics. *Nuclear Physics* **28**, 558 (1978).

RECURRENCE RELATIONS FOR THE COEFFICIENTS IN HYPERGEOMETRIC SERIES EXPANSIONS

L. REBILLARD

Symbolic Computation Group, University of Waterloo,
Ontario, Canada, N2L 3G1
E-mail: lrebillard@scg.uwaterloo.ca

H. ZAKRAJŠEK

Faculty of Mechanical Engineering, University of Ljubljana,
Aškerčeva 6, 1000 Ljubljana, Slovenia
E-mail: helena.zakrajsek@fs.uni-lj.si

Considering an arbitrary family of hypergeometric polynomials $\{P_n\}$ and a linear differential operator with polynomial coefficients L, we present an algorithm, based on the manipulation of hypergeometric families, that generates a recurrence relation for the coefficient sequence $\{c_n\}$ satisfying $L(\sum c_n P_n(x)) = 0$.

1. Introduction

Let $\{P_n\}_{n\in\mathbb{N}}$ be a family of hypergeometric polynomials (HPF) and let f be a function satisfying a linear differential equation with polynomial coefficients

$$\sum_{k=0}^{r} p_k(x)D^k f(x) = q(x), \tag{1}$$

where $D := d/dx$ is the ordinary derivative operator. We are looking for a recurrence relation

$$\mathcal{L}(c_n) = q_n, \quad \forall n \geq 0, \tag{2}$$

satisfied by the sequence of coefficients $c = (c_n)_{n\geq 0}$ in a formal series expansion

$$f(x) = \sum_{n=0}^{\infty} c_n P_n(x), \tag{3}$$

where the sequence $(q_n)_{n \geq 0}$ depends only on the right-hand side $q(x)$. If the family $\{P_n\}$ is orthogonal on the support I with respect to a weight function ρ, the coefficient c_n is the n-th Fourier coefficient of f in the expansion with respect to the $\{P_n\}$ basis.

Some well-known special cases of this problem are:

(1) *connection problems*, where $f := Q_m$, $m \in \mathbb{N}$, and $\{Q_m\}$ is another (classical orthogonal) polynomial family or a classical associated polynomial family; in particular, *inversion problems*, where $f := x^m$ (see Refs. 4,10–12,21,22)

(2) *linearization problems*, where $f := \prod_{j=1}^m P_{k_j}$, $m, k_j \in \mathbb{N}$, and in particular, $f := P_k^m$ (see Refs. 4,13,22)

(3) *solutions of differential equations by τ-methods* (see Refs. 5,6,17)

Recently, different methods for the construction of a recurrence relation have been developed. Lewanowicz' algorithm[7–13] is applicable to any classical orthogonal family and uses certain identities involving the Fourier coefficients of a function f to construct a recurrence relation for the coefficients. Another method was proposed by the NAVIMA group[21,22] where orthogonality is not required, however the recurrence relation obtained by this method is not always minimal.

A third approach based on the manipulation of formal series of hypergeometric polynomials has been presented in Refs. 19 and 20 and implemented in the package ORTHOGONALSERIES of the computer algebra system MAPLE[14] (since version MAPLE7). Given a formal series expanded with respect to an arbitrary family $\{P_n\}$ of hypergeometric polynomials and a linear differential operator L with polynomial coefficients of order r, one can build a recurrence operator $\mathcal{L}$ such that

$$L\left(\sum_{n=0}^{\infty} c_n P_n\right) = \sum_{n=0}^{\infty} \mathcal{L}(c_n) P_n^{(r)} \tag{4}$$

where $\{P_n^{(r)}\}$ is a family of hypergeometric polynomials proportional to the r-th derivatives of the P_n. And by expanding - when possible - the function q as

$$q = \sum_{n=0}^{\infty} q_n P_n^{(r)} \tag{5}$$

we obtain a recurrence through term-by-term identification. In this paper, we improve the process for applying differential operators on a series of hypergeometric polynomials presented in Refs. 19,20 and obtain the following general formula

$$L\left(\sum_{n=0}^{\infty} c_n P_n\right) = \sum_{n=0}^{\infty} \widetilde{\mathcal{L}}(c_n) P_n^{(i,j)} \qquad (6)$$

where $\{P_n^{(i,j)}\}$, with suitable $i, j \in \mathbb{Z}$, is a HPF related to the original one. In (6) the order of the operator $\widetilde{\mathcal{L}}$ is smaller or equal to the order of $\mathcal{L}$ if the expanding basis $\{P_n^{(i,j)}\}$ is cleverly chosen. Hence we build a recurrence $\widetilde{\mathcal{L}}(c_n) = \tilde{q}_n$ where

$$q = \sum_{n=0}^{\infty} \tilde{q}_n P_n^{(i,j)}. \qquad (7)$$

Remark 1.1. In fact, legitimacy of this identification requires normal convergence results for series appearing in (4) and (7), and independency (orthogonality implies it) of the family $\{P_n^{(r)}\}$. However, in most applications (see above) the right-hand side q is a polynomial (often zero) and the searched solution of (1) is also a polynomial, so that the expansion (3) exists, no convergence problems occur and identification does not require orthogonality. For this reason, in the rest of this paper, we deal with formal series without consideration of convergence and with generic HPF without consideration of orthogonality.

The structure of the paper is as follows: In Section 2 we introduce notations and briefly recall the basic properties of hypergeometric families. We also present the method implemented in the package ORTHOGONALSERIES in more details as it was introduced in an unpublished PhD Thesis.[19] In Section 3 we introduce special HPFs $\{P_n^{(i,j)}\}$ associated to $\{P_n\}$ family and define differential and recurrence operators relating P_n and $P_n^{(i,j)}$. In Section 4 we show that different expansion basis $\{P_n^{(i,j)}\}$ can be used and how the order of the operator $\mathcal{L}$ depends on a chosen basis. We also establish which expansion basis (we call it the minimal expansion basis and it is uniquely determined) has to be used to compute $\mathcal{L}$ of the lowest order. In Section 5 we describe the algorithm that rewrites the differential operator in a special form (depending on the given hypergeometric family) suitable for the construction of a recurrence relation satisfied by the coefficient sequence of the hypergeometric series. The derived recurrence is sometimes of lower order than the one obtained by Lewanowicz' method (see Example 5.1).

2. Notations and basic properties

Let σ and τ be polynomials of degrees not higher than two and one, respectively, and let us associate to them the values

$$\lambda_n = -n\left(\tau' + (n-1)\frac{\sigma''}{2}\right).$$

If the function $n \mapsto \lambda_n$ is injective for positive integers, then for all $n \in \mathbb{N}$ ($\mathbb{N}$ denotes the set of nonnegative integers) the *hypergeometric differential equation*

$$\sigma(x)y''(x) + \tau(x)y'(x) + \lambda_n y(x) = 0 \tag{8}$$

has a polynomial solution P_n of degree exactly n. The polynomials P_n are called *hypergeometric polynomials* and are determined by the Rodrigues' type formula[16]

$$P_n(x) = \frac{B_n}{\rho(x)}D^n[\rho(x)\sigma^n(x)],$$

where B_n is an arbitrary normalization coefficient and the function ρ is a solution of the Pearson differential equation

$$(\sigma\rho)' = \tau\rho. \tag{9}$$

Therefore a *hypergeometric family* $\{P_n\}$ is entirely determined by the triplet $[\sigma, \tau, B_n]$, which will be denoted by

$$P_n \longleftarrow [\sigma, \tau, B_n].$$

Every HPF satisfies (see Refs. 15,16,19) the *three-term recurrence relation*, the *structure relation* and the *derivative representation*, respectively,

$$x\Gamma_n(x) - x_1(n)\Gamma_{n+1}(x) + x_0(n)P_n(x) + x_{-1}(n)P_{n-1}(x), \tag{10}$$

$$\sigma(x)DP_n(x) = \sigma_1(n)P_{n+1}(x) + \sigma_0(n)P_n(x) + \sigma_{-1}(n)P_{n-1}(x), \tag{11}$$

$$P_n(x) = \rho_1(n)DP_{n+1}(x) + \rho_0(n)DP_n(x) + \rho_{-1}(n)DP_{n-1}(x), \tag{12}$$

where we conventionally assume that $P_{-1} = 0$. Coefficients x_i, σ_i, ρ_i are functions of n and depend only on σ, τ and B_n (see Refs. 15,16,19). In particular, these functions are rational in n if B_n is hypergeometric. Let us introduce the notation

$$\tau_n = \tau + n\sigma',$$

$$\kappa_n = \tau'_{\frac{n-1}{2}},$$

$$x_n^* = -\frac{\tau_n(0)}{\tau'_n}.$$

Then we have the explicit formulas for coefficients x_i, σ_i, ρ_i:

$$x_1(n) = \begin{cases} \dfrac{B_n}{B_{n+1}} \dfrac{\kappa_n}{\kappa_{2n}\tau_n'}, & n > 0, \\[2ex] \dfrac{B_0}{B_1\tau'}, & n = 0, \end{cases}$$

$$x_0(n) = \begin{cases} n\dfrac{\tau_{n-1}(0)}{\tau_{n-1}'} - (n+1)\dfrac{\tau_n(0)}{\tau_n'}, & n > 0, \\[2ex] -\dfrac{\tau(0)}{\tau'}, & n = 0, \end{cases}$$

$$x_{-1}(n) = \begin{cases} -\dfrac{nB_n}{B_{n-1}} \dfrac{\tau_{n-1}'\sigma(x_{n-1}^*)}{\kappa_{2n}}, & n > 0, \\[2ex] 0, & n = 0, \end{cases}$$

$$\begin{aligned}
\sigma_1(n) &= \tfrac{n}{2}\sigma'' x_1(n), & \rho_1(n) &= \tfrac{1}{n+1}x_1(n), \\
\sigma_0(n) &= -\tfrac{\kappa_n}{\tau_n'}\tau_n(x_0(n)) = -\tfrac{1}{2}\tau(x_0(n)), & \rho_0(n) &= -\tfrac{1}{2\lambda_n}\tau(x_0(n)), \\
\sigma_{-1}(n) &= -\kappa_n x_{-1}(n), & \rho_{-1}(n) &= -\tfrac{\sigma''}{2\kappa_{n-1}}x_{-1}(n).
\end{aligned}$$

Let E denote the shift operator acting on sequences $\{x_n\}_{n\geq 0}$ by $Ex_n = x_{n+1}$. Define E^{-1} by $E^{-1}x_n = x_{n-1}$ for $n \geq 1$ and $E^{-1}x_0 = 0$. By introducing recurrence operators

$$\begin{aligned}
\mathbf{X} &= x_1(n)E + x_0(n) + x_{-1}(n)E^{-1}, \\
\mathbf{S} &= \sigma_1(n)E + \sigma_0(n) + \sigma_{-1}(n)E^{-1}, \\
\mathbf{R} &= \rho_1(n) + \rho_0(n)E^{-1} + \rho_{-1}(n)E^{-2},
\end{aligned}$$

we can rewrite (10) – (12) in a more compact form

$$xP_n = \mathbf{X}P_n, \tag{13}$$

$$\sigma(x)DP_n = \mathbf{S}P_n, \tag{14}$$

$$P_n = \mathbf{R}DP_{n+1}. \tag{15}$$

The derivatives of hypergeometric polynomials are still hypergeometric,[16] namely DP_{n+1} is a solution of

$$\sigma(x)y''(x) + [\tau(x) + \sigma'(x)]y'(x) + [\lambda_{n+1} + \tau'(x)]y(x) = 0,$$

and is characterized by

$$DP_{n+1}(x) = -\frac{\lambda_{n+1}B_{n+1}}{\sigma(x)\rho(x)}D^n[\rho(x)\sigma^{n+1}(x)], \qquad (16)$$

hence

$$DP_{n+1} \longleftarrow [\sigma, \tau + \sigma', -\lambda_{n+1}B_{n+1}]. \qquad (17)$$

More generally, the family $\{D^k P_{n+k}\}$ is a family of hypergeometric polynomials and it can be easily shown that

$$D^k P_{n+k} \longleftarrow [\sigma, \tau_k, B_n^{(k)}]$$

with a suitable normalization constant $B_n^{(k)}$. Of course, with any other (non zero) normalization coefficient we obtain a family of hypergeometric polynomials proportional to the $D^k P_{n+k}$ that satisfies

$$\sigma(x)y''(x) + \tau_k(x)y'(x) + \lambda_n^{(k)}y(x) = 0 \qquad (18)$$

with

$$\lambda_n^{(k)} = -n\left(\tau_k' + (n-1)\frac{\sigma''}{2}\right).$$

For convenience and consistency of notations in the rest of the paper we introduce the polynomial $P_n^{(k)}$ of degree n defined by

$$P_n^{(k)} \longleftarrow [\sigma, \tau_k, B_n]. \qquad (19)$$

It can be easily verified that

$$P_n^{(0)} = P_n, \qquad (20)$$

$$P_n^{(1)} = -\frac{B_n}{\lambda_{n+1}B_{n+1}}DP_{n+1}, \qquad (21)$$

$$P_n^{(k)} = -\frac{B_n}{\lambda_{n+1}^{(k-1)}B_{n+1}}DP_{n+1}^{(k-1)}. \qquad (22)$$

Applying the results presented for P_n at the beginning of this section, for each positive integer k there exist recurrence operators $\mathbf{X}^{(k)}$, $\mathbf{S}^{(k)}$, $\mathbf{R}^{(k)}$ and $\mathbf{D}^{(k)}$ such that

$$xP_n^{(k)} = \mathbf{X}^{(k)}P_n^{(k)}, \qquad (23)$$

$$\sigma(x)DP_n^{(k)} = \mathbf{S}^{(k)}P_n^{(k)}, \qquad (24)$$

$$P_n^{(k)} = \mathbf{R}^{(k)}P_n^{(k+1)}, \qquad (25)$$

$$DP_n^{(k)} = \mathbf{D}^{(k)}P_n^{(k+1)}, \qquad (26)$$

where

$$\mathbf{D}^{(k)} = -\frac{\lambda_n^{(k)} B_n}{B_{n-1}} E^{-1}.$$

In particular, for $k = 0$, $\mathbf{X}^{(0)} = \mathbf{X}$, $\mathbf{S}^{(0)} = \mathbf{S}$ and $\mathbf{R}^{(0)} = -\mathbf{R}\lambda_{n+1} B_{n+1}/B_n$. At this stage, we have a set of formulas that allow us to perform some elementary operations on hypergeometric polynomials. For a given differential operator L of order r with polynomial coefficients we are now able to find a corresponding recurrence operator $\mathbf{L}$ such that $L P_n = \mathbf{L} P_n^{(r)}$ and the adjoint operator enables the construction of a recurrence relation for the coefficient sequence that we are looking for.

Let us recall that for a recurrence operator $\mathbf{L}$

$$\mathbf{L} = \sum_{k=r}^{s} l_k(n) E^k, \quad r, s \in \mathbb{Z},$$

its *adjoint operator* $\mathbf{L}^*$ is defined by (see Refs. 1,18,19)

$$\mathbf{L}^* = \sum_{k=-s}^{-r} l_{-k}(n+k) E^k.$$

Note that $\mathbf{L}^{**} = \mathbf{L}$ and $(\mathbf{L}\mathbf{M})^* = \mathbf{M}^* \mathbf{L}^*$. For brevity and clearness in presence of superscripts we use script style letters to denote adjoint operators. Therefore the adjoint operator of a generic recurrence operator $\mathbf{L}$ will be denoted by $\mathcal{L} := \mathbf{L}^*$. Having two sequences of functions $(a_n)_{n\in\mathbb{Z}}$ and $(b_n)_{n\in\mathbb{Z}}$ it can be readily checked that

$$\sum_{n\in\mathbb{Z}} a_n \mathbf{L}(b_n) = \sum_{n\in\mathbb{Z}} \mathcal{L}(a_n) b_n, \tag{27}$$

where the two sums are treated as formal series.

Therefore for a differential operator L with polynomial coefficients p_k

$$L = \sum_{k=0}^{r} p_k(x) D^k, \tag{28}$$

there exists a recurrence operator $\mathbf{L}$ and its adjoint operator $\mathcal{L}$ such that

$$L\left(\sum_{n=0}^{\infty} c_n P_n\right) = \sum_{n=0}^{\infty} c_n \mathbf{L} P_n^{(r)} = \sum_{n=0}^{\infty} \mathcal{L}(c_n) P_n^{(r)}. \tag{29}$$

Indeed,

$$LP_n = \sum_{k=0}^{r} p_k(x) D^k P_n = \sum_{k=0}^{r} p_k(x) \mathbf{D}^{(0)} \mathbf{D}^{(1)} \cdots \mathbf{D}^{(k-1)} P_n^{(k)}$$

$$= \sum_{k=0}^{r} p_k(x) \mathbf{D}^{(0)} \mathbf{D}^{(1)} \cdots \mathbf{D}^{(k-1)} \mathbf{R}^{(k)} \cdots \mathbf{R}^{(r-1)} P_n^{(r)}$$

$$= \sum_{k=0}^{r} \mathbf{D}^{(0)} \mathbf{D}^{(1)} \cdots \mathbf{D}^{(k-1)} \mathbf{R}^{(k)} \cdots \mathbf{R}^{(r-1)} p_k(\mathbf{X}^{(r)}) P_n^{(r)}$$

hence the result follows

$$\mathcal{L} = \sum_{k=0}^{r} p_k(\mathcal{X}^{(r)}) \mathcal{R}^{(r-1)} \cdots \mathcal{R}^{(k)} \mathcal{D}^{(k-1)} \cdots \mathcal{D}^{(0)}. \tag{30}$$

This is the method used in the ORTHOGONALSERIES package. A drawback of this straightforward and simple procedure is that the resulting recurrence operator $\mathcal{L}$ is not always of minimal order. This happens, for example, when the leading coefficient p_r of L has roots in common with σ. We illustrate this fact hereafter.

Example 2.1. If $L = \sigma(x)D + x$ then by (30) we obtain $\mathcal{L} = \sigma(\mathcal{X}^{(1)})\mathcal{D}^{(0)} + \mathcal{X}^{(1)}\mathcal{R}^{(0)}$ of order 4. Note that the expansion basis in this case is $\{P_n^{(1)}\}$. However, if we consider the structure relation (14) we obtain the operator $\widetilde{\mathcal{L}} = \mathcal{S} + \mathcal{X}$ of order 2 due to the fact that the expansion basis is now $\{P_n\}$. It can be computed that $\mathcal{L} = \mathcal{R}^{(0)}\widetilde{\mathcal{L}}$. Let us illustrate this remark in the case of Laguerre polynomials with parameter α. Hence the result of the operation

$$[\tau D + \tau] \sum_{n=1}^{\infty} u_n L_n^{(\alpha)}(x)$$

using the general ORTHOGONALSERIES process is

$$C_0 L_0^{(\alpha+1)}(x) + \sum_{n=1}^{\infty} C_n L_n^{(\alpha+1)}(x)$$

with

$$C_0 = (2 + \alpha)u_0 - (6 + 3\alpha)u_1 + (4 + 2\alpha)u_2$$
$$C_n = (\alpha + 2 + 4n)u_n - (6 + 5n + 3\alpha)u_{n+1} - nu_{n-1} + (2\alpha + 4 + 2n)u_{n+2}$$

whereas the result obtained by taking account of the structural derivation is

$$D_0 L_0^{(\alpha)}(x) + \sum_{n=1}^{\infty} D_n L_n^{(\alpha)}(x)$$

with

$$D_0 = (1 + \alpha)u_0 - (2\alpha + 2)u_1$$
$$D_n = (3n + \alpha + 1)u_n - nu_{n-1} - (2n + 2\alpha + 2)u_{n+1}.$$

In the rest of the paper we show how the ORTHOGONALSERIES method can be improved in order to lower the order of resulting operators. In particular, it will be proved that, as shown in the previous example, the order of the constructed recurrence operator depends strongly on the expansion basis $\{P_n^{(k)}\}$ used. The lower k is, the lower the order of the resulting operator is. Namely, if $LP_n = \mathbf{L}_1 P_n^{(k)}$ and $LP_n = \mathbf{L}_2 P_n^{(k+1)}$, then by (25) $\mathbf{L}_1 \mathbf{R}^{(k)} = \mathbf{L}_2$. So $\mathcal{L}_2 = \mathcal{R}^{(k)} \mathcal{L}_1$ and $\mathcal{L}_1$ has lower order than $\mathcal{L}_2$. This indicates another theoretically interesting approach to recurrence construction. Since we know which are possible redundant factors on the left (operators $\mathbf{R}^{(k)}$), one can try to factor the recurrence obtained by the general process. However, from the computational point of view this method has a defect. In general, the order of the recurrence obtained by the general process is large and, especially when working with symbolic parameters (e.g. in the case of Jacobi family $\{P_n^{(\alpha,\beta)}\}$), the factorization attempt is very time and space consuming.

3. Associated families

In this section we introduce new families of hypergeometric polynomials related to $\{P_n\}$ which will be used in the improved algorithm to lower the order of a recurrence relation whenever possible. Let σ be a polynomial of degree 2 and let ξ_1 and ξ_2 be the roots of σ (not necessarily distinct).

Definition 3.1. Let $\{P_n\}$ be a HPF determined by the triplet $[\sigma, \tau, B_n]$. For each pair (i, j) of integers we introduce the associated family $\{P_n^{(i,j)}\}$ of hypergeometric polynomials determined by

$$P_n^{(i,j)} \longleftarrow [\sigma, \tau + (i(x - \xi_1) + j(x - \xi_2))\sigma''/2, B_n].$$

In particular,

$$P_n^{(0,0)} = P_n,$$
$$P_n^{(1,0)} \longleftarrow [\sigma, \tau + (x - \xi_1)\sigma''/2, B_n],$$
$$P_n^{(0,1)} \longleftarrow [\sigma, \tau + (x - \xi_2)\sigma''/2, B_n],$$
$$P_n^{(1,1)} \longleftarrow [\sigma, \tau + \sigma', B_n].$$

Here we must verify if the polynomials $P_n^{(i,j)}$ are well defined. In section 2 we mentioned that the polynomials P_n are well defined if the function

$n \to \lambda_n$ is injective on $\mathbb{N}$. As the degree of σ is 2, this is equivalent to

$$\frac{-2\tau'}{\sigma''} \notin \mathbb{N}.$$

Therefore, $P_n^{(i,j)}$ is well-defined if

$$-2\frac{\tau' + \frac{\sigma''}{2}(i+j)}{\sigma''} = \frac{-2\tau'}{\sigma''} - i - j \notin \mathbb{N}.$$

Then, if the polynomials P_n are well defined then two cases can occur

(1) $\frac{-2\tau'}{\sigma''} \notin \mathbb{Z}$ then polynomials $P_n^{(i,j)}$ exist for any i and j in $\mathbb{Z}$.
(2) $\frac{-2\tau'}{\sigma''} = k \in \mathbb{Z} - \mathbb{N}$ then polynomials $P_n^{(i,j)}$ exist for any i and j in $\mathbb{Z}$ such that $i + j > k$.

Without a big loss of generality, we will assume in the rest of this article that any polynomial $P_n^{(i,j)}$ appearing in further results is well-defined. Note that, because of the identity $\sigma' = \frac{\sigma''}{2}[(x - \xi_1) + (x - \xi_2)]$, we have $P_n^{(k,k)} = P_n^{(k)}$ as in (19) and by (21) we have

$$P_n^{(1,1)} = -\frac{B_n}{\lambda_{n+1} B_{n+1}} D P_{n+1}. \tag{31}$$

Definition 3.2. Let $\sigma = \sigma''(x - \xi_1)(x - \xi_2)/2$ and let $\{P_n\}$ be a HPF determined by the triplet $[\sigma, \tau, B_n]$. For $i, j \in \mathbb{Z}$ we define differential operators

$$U^{(i,j)} = \sigma D + \tau + \frac{\sigma''}{2}[(i-1)(x - \xi_1) + (j-1)(x - \xi_2)],$$

$$T_1^{(i,j)} = (x - \xi_1)D + \frac{\tau(\xi_1)}{\sigma'(\xi_1)} + j - 1, \qquad (\text{if } \xi_1 \neq \xi_2),$$

$$T_2^{(i,j)} = (x - \xi_2)D + \frac{\tau(\xi_2)}{\sigma'(\xi_2)} + i - 1, \qquad (\text{if } \xi_1 \neq \xi_2),$$

and recurrence operators

$$\mathbf{D}^{(i,j)} = \frac{nB_n}{B_{n-1}}\kappa_{n+i+j}E^{-1},$$

$$\mathbf{T}_1^{(i,j)} = n + \frac{\tau(\xi_1)}{\sigma'(\xi_1)} + j, \qquad (\text{if } \xi_1 \neq \xi_2),$$

$$\mathbf{T}_2^{(i,j)} = n + \frac{\tau(\xi_2)}{\sigma'(\xi_2)} + i, \qquad (\text{if } \xi_1 \neq \xi_2),$$

$$\mathbf{S}_1^{(i,j)} = \frac{1}{\kappa_{2n+i+j}}\left(\frac{B_n}{B_{n+1}}E - \tau_{n-1+j}(\xi_1)\right),$$

$$\mathbf{S}_2^{(i,j)} = \frac{1}{\kappa_{2n+i+j}}\left(\frac{B_n}{B_{n+1}}E - \tau_{n-1+i}(\xi_2)\right),$$

$$\mathbf{R}_1^{(i,j)} = \frac{1}{\kappa_{2+i+j}}\left(\kappa_{n+i+j} + \frac{\sigma''}{2}\frac{nB_n}{B_{n-1}}\tau_{n-1+j}(\xi_1)E^{-1}\right),$$

$$\mathbf{R}_2^{(i,j)} = \frac{1}{\kappa_{2n+i+j}}\left(\kappa_{n+i+j} + \frac{\sigma''}{2}\frac{nB_n}{B_{n-1}}\tau_{n-1+i}(\xi_2)E^{-1}\right).$$

In the rest of the section we show how these operators relate the elements of different associated families. As all the statements appear in symmetrical pairs (one for each root of σ), we only prove one of them.

Lemma 3.1.

$$P_n^{(1,0)} = P_n + \frac{nB_n}{B_{n-1}}\frac{\sigma''}{2}(x - \xi_1)P_{n-1}^{(1,1)}, \tag{32}$$

$$P_n^{(0,1)} = P_n + \frac{nB_n}{B_{n-1}}\frac{\sigma''}{2}(x - \xi_2)P_{n-1}^{(1,1)}. \tag{33}$$

Proof. Let ρ^* be a solution of the Pearson differential equation associated with $P_n^{(1,0)}$. By (9)

$$(\sigma\rho^*)' = \left(\tau + \frac{\sigma''}{2}(x - \xi_1)\right)\rho^* = \left(\frac{\tau}{\sigma} + \frac{1}{x - \xi_2}\right)\sigma\rho^*,$$

hence $\sigma\rho^* = \text{const}(x - \xi_2)\exp\left(\int \tau/\sigma\,dx\right)$ and we obtain $\rho^* = (x - \xi_2)\rho$. Using the Rodrigues' type formula for $P_n^{(1,0)}$ it follows that

$$
\begin{aligned}
P_n^{(1,0)} &= \frac{B_n}{(x - \xi_2)\rho}D^n[(x - \xi_2)\rho\sigma^n] \\
&= \frac{B_n}{(x - \xi_2)\rho}[(x - \xi_2)D^n(\rho\sigma^n) + nD^{n-1}(\rho\sigma^n)] \\
&= P_n - \frac{\sigma n}{(x - \xi_2)\lambda_n}P_n' \\
&= P_n + \frac{nB_n}{B_{n-1}}\frac{\sigma''}{2}(x - \xi_1)P_{n-1}^{(1,1)},
\end{aligned}
$$

where we used (16) and (31) in the last two lines. $\qquad\square$

Remark 3.1. If the polynomials P_n are orthogonal with respect to the weight ρ (solution of the Pearson equation) then $P_n^{(1,0)}$ (resp. $P_n^{(0,1)}$) is the **kernel polynomial** $P_n^*(\xi_2; x)$ (resp. $P_n^*(\xi_1; x)$) using definition and notation introduced by Chihara[2] in section 7 of chapter 1.

Lemma 3.2.

$$
\begin{aligned}
(x - \xi_1)P_n^{(i,j)} &= \mathbf{S}_1^{(i,j)}P_n^{(i,j-1)}, && (34) \\
(x - \xi_2)P_n^{(i,j)} &= \mathbf{S}_2^{(i,j)}P_n^{(i-1,j)}. && (35)
\end{aligned}
$$

Proof. In view of (31), (13) and (14) multiplying (33) by $(x - \xi_1)$ results in

$$
(x - \xi_1)P_n^{(0,1)} = (x - \xi_1)P_n - \frac{n}{\lambda_n}\sigma D P_n = (\mathbf{X} - \xi_1 - \frac{n}{\lambda_n}\mathbf{S})P_n.
$$

Calculating (note that $\tau_n(x) = \tau_n' \cdot x + \tau_n(0)$)

$$
\begin{aligned}
\mathbf{X} - \xi_1 - \frac{n}{\lambda_n}\mathbf{S} &= \left(1 - \frac{n^2\sigma''}{2\lambda_n}\right)x_1(n)E + \left(x_0(n) - \xi_1 - \frac{\tau_n(x_0(n))}{\tau_n'}\right) \\
&= \frac{B_n}{B_{n+1}\tau_n'}E - \frac{\tau_n(\xi_1)}{\tau_n'}
\end{aligned}
$$

we obtain

$$
(x - \xi_1)P_n^{(0,1)} = \left(\frac{B_n}{B_{n+1}\tau_n'}E - \frac{\tau_n(\xi_1)}{\tau_n'}\right)P_n.
$$

Now replace P_n by $P_n^{(i,j-1)}$ which means that $P_n^{(0,1)}$, τ, τ_n' and $\tau_n(\xi_1)$ are replaced by $P_n^{(i,j)}$, $\tau + i(x - \xi_1)\sigma''/2 + (j - 1)(x - \xi_2)\sigma''/2$, $\tau_{n+(i+j-1)/2}'$

and $\tau_{n+j-1}(\xi_1)$, respectively. Therefore

$$(x - \xi_1)P_n^{(i,j)} = \left(\frac{B_n}{B_{n+1}\tau'_{n+(i+j-1)/2}} E - \frac{\tau_{n+j-1}(\xi_1)}{\tau'_{n+(i+j-1)/2}} \right) P_n^{(i,j-1)}. \qquad \square$$

Lemma 3.3.

$$P_n^{(i,j)} = \mathbf{R}_1^{(i,j)} P_n^{(i+1,j)}, \tag{36}$$
$$P_n^{(i,j)} = \mathbf{R}_2^{(i,j)} P_n^{(i,j+1)}. \tag{37}$$

Proof. By Lemma 3.2, for $i = j = 1$ we have

$$(x - \xi_1)P_{n-1}^{(1,1)} = \frac{B_{n-1}}{B_n\tau'_{n-1/2}} P_n^{(1,0)} - \frac{\tau_{n-1}(\xi_1)}{\tau'_{n-1/2}} P_{n-1}^{(1,0)}.$$

Combining this with Lemma 3.1 we obtain

$$P_n = P_n^{(1,0)} - \frac{nB_n}{B_{n-1}} \frac{\sigma''}{2} \left[\frac{B_{n-1}}{B_n\tau'_{n-1/2}} P_n^{(1,0)} - \frac{\tau_{n-1}(\xi_1)}{\tau'_{n-1/2}} P_{n-1}^{(1,0)} \right]$$

$$= \left(1 - \frac{n\sigma''}{2\tau'_{n-1/2}} \right) P_n^{(1,0)} + \frac{nB_n}{B_{n-1}} \frac{\sigma''}{2} \frac{\tau_{n-1}(\xi_1)}{\tau'_{n-1/2}} P_{n-1}^{(1,0)}.$$

Note that $2\tau'_{n-1/2} - n\sigma'' = 2\tau'_{(n-1)/2}$. Replace P_n by $P_n^{(i,j)}$, hence $P_n^{(1,0)}$, $\tau_{n-1}(\xi_1)$ and τ'_n are replaced by $P_n^{(i+1,j)}$, $\tau_{n-1+j}(\xi_1)$ and $\tau'_{n+(i+j)/2}$, respectively, and the first statement is proved. $\qquad \square$

Proposition 3.1. *For HPF $\{P_n\}$ determined by $[\sigma, \tau, B_n]$ the following are true:*

$$DP_n^{(i-1,j-1)} = \mathbf{D}^{(i-1,j-1)} P_n^{(i,j)}, \tag{38}$$
$$U^{(i,j)} P_n^{(i,j)} = \frac{B_n}{B_{n+1}} EP_n^{(i-1,j-1)}. \tag{39}$$

Proof. Let $\tilde{\tau} = \tau + \frac{\sigma''}{2}[(i-1)(x-\xi_1) + (j-1)(x-\xi_2)]$. Then we have

$$P_n^{(i-1,j-1)} \longleftarrow [\sigma, \tilde{\tau}, B_n] \qquad \text{and} \qquad P_n^{(i,j)} \longleftarrow [\sigma, \tilde{\tau} + \sigma', B_n].$$

In view of (17) we obtain

$$DP_{n+1}^{(i-1,j-1)} \longleftarrow [\sigma, \tilde{\tau} + \sigma', (n+1)\tilde{\tau}'_{n/2}B_{n+1}],$$

hence replacing n by $n - 1$ results in

$$DP_n^{(i-1,j-1)} = \frac{nB_n}{B_{n-1}} \tau'_{(n+i+j-3)/2} E^{-1} P_n^{(i,j)} = \mathbf{D}^{(i-1,j-1)} P_n^{(i,j)}.$$

As $P_n^{(i-1,j-1)}$ satisfies a hypergeometric differential equation

$$(\sigma D^2 + \tilde{\tau} D + \lambda_n^{(i-1,j-1)})y = 0,$$

it follows that $U^{(i,j)}DP_n^{(i-1,j-1)} = n\kappa_{n+i+j-2}P_n^{(i-1,j-1)}$, hence by (38) we obtain (39). $\square$

Lemma 3.4.

$$P_n = P_n^{(1,-1)} + \frac{nB_n}{B_{n-1}}\sigma'(\xi_1)P_{n-1}^{(1,0)},$$

$$P_n = P_n^{(-1,1)} + \frac{nB_n}{B_{n-1}}\sigma'(\xi_2)P_{n-1}^{(0,1)}.$$

Proof. To prove the first statement, we write (36) for $i = 0, j = 0$ and (37) for $i = 1, j = -1$, and subtract

$$P_n - P_n^{(1,-1)} = (\mathbf{R}_1^{(0,0)} - \mathbf{R}_2^{(1,-1)})P_n^{(1,0)}$$

$$= \frac{nB_n}{B_{n-1}}\frac{\sigma''}{2}\frac{\tau_{n-1}(\xi_1) - \tau_n(\xi_2)}{\tau'_{n-1/2}}P_{n-1}^{(1,0)}.$$

As $\tau_{n-1}(\xi_1) - \tau_n(\xi_2) = (\xi_1 - \xi_2)\tau' + (2n-1)\sigma'(\xi_1)$ and $(\xi_1 - \xi_2) = 2\sigma'(\xi_1)/\sigma''$, the proof is concluded. $\square$

Lemma 3.5.

$$(x - \xi_1)DP_n = nP_n - \frac{nB_n}{B_{n-1}}\tau_{n-1}(\xi_1)P_{n-1}^{(1,0)},$$

$$(x - \xi_2)DP_n = nP_n - \frac{nB_n}{B_{n-1}}\tau_{n-1}(\xi_2)P_{n-1}^{(0,1)}.$$

Proof. Put $i = j = 1$ in (34) and keep in mind (38). Then we subtract (36) (put $i = j = 0$) multiplied by n and the result follows. $\square$

Lemma 3.6. *Let $\xi_1 \neq \xi_2$ be distinct roots of σ. Then*

$$T_1^{(0,0)}P_n = \mathbf{T}_1^{(0,0)}P_n^{(1,-1)},$$

$$T_2^{(0,0)}P_n = \mathbf{T}_2^{(0,0)}P_n^{(-1,1)}.$$

Proof. In view of Lemma 3.5 and Lemma 3.4 we write

$$T_1^{(0,0)} P_n =$$

$$= nP_n - \frac{nB_n}{B_{n-1}}\tau_{n-1}(\xi_1)P_{n-1}^{(1,0)} + \left(\frac{\tau(\xi_1)}{\sigma'(\xi_1)} - 1\right)\left[P_n^{(1,-1)} + \frac{nB_n}{B_{n-1}}\sigma'(\xi_1)P_{n-1}^{(1,0)}\right]$$

$$= \left(\frac{\tau(\xi_1)}{\sigma'(\xi_1)} - 1\right)P_n^{(1,-1)} + n\left[P_n - \frac{B_n}{B_{n-1}}(\tau_{n-1}(\xi_1) - \tau(\xi_1) + \sigma'(\xi_1))P_{n-1}^{(1,0)}\right]$$

$$= \left(\frac{\tau(\xi_1)}{\sigma'(\xi_1)} - 1\right)P_n^{(1,-1)} + n\left[P_n - \frac{B_n}{B_{n-1}}n\sigma'(\xi_1))P_{n-1}^{(1,0)}\right]$$

$$= \left(n + \frac{\tau(\xi_1)}{\sigma'(\xi_1)} - 1\right)P_n^{(1,-1)}. \qquad \square$$

Proposition 3.2. *Let ξ_1 and ξ_2 be distinct roots of σ. Then*

$$T_1^{(i,j)} P_n^{(i,j)} = \mathbf{T}_1^{(i,j)} P_n^{(i+1,j-1)},$$
$$T_2^{(i,j)} P_n^{(i,j)} = \mathbf{T}_2^{(i,j)} P_n^{(i-1,j+1)}.$$

Proof. As $\tau^{(i,j)}(\xi_1) = \tau(\xi_1) + j\frac{\sigma''}{2}(\xi_1-\xi_2) = \tau(\xi_1) + j\sigma'(\xi_1)$ and $\tau^{(i,j)}(\xi_2) = \tau(\xi_2) + i\frac{\sigma''}{2}(\xi_2-\xi_1) = \tau(\xi_2) + i\sigma'(\xi_2)$ in view of Lemma 3.6 the result follows.

Example 3.1. The Jacobi polynomials $P_n^{(\alpha,\beta)}$ are determined by the triplet $[1 - x^2, \beta - \alpha - (\alpha + \beta + 2)x, (-1)^n/(2^n n!)]$. Moreover, the associated family with parameters (i,j) is still a family of Jacobi polynomials $P_n^{(\alpha+i,\beta+j)}$ as we can check from its defining triplet (we put $\xi_1 = -1, \xi_2 = 1$)

$$\left[1 - x^2, \beta - \alpha - i + j - (\alpha + \beta + 2 + i + j)x, (-1)^n/(2^n n!)\right].$$

Therefore, by Proposition 3.2 and

$$DP_n^{(\alpha,\beta)} = \frac{n + \alpha + \beta + 1}{2}P_{n-1}^{(\alpha+1,\beta-1)}$$

we obtain two (apparently new) recurrence relations between Jacobi polynomials

$$\frac{n + \alpha + \beta + 1}{2}(x + 1)P_{n-1}^{(\alpha+1,\beta+1)} + \beta P_n^{(\alpha,\beta)} = (n + \beta)P_n^{(\alpha+1,\beta-1)},$$

$$\frac{n + \alpha + \beta + 1}{2}(x - 1)P_{n-1}^{(\alpha+1,\beta+1)} + \alpha P_n^{(\alpha,\beta)} = (n + \alpha)P_n^{(\alpha-1,\beta+1)}.$$

4. Depression of the order

As mentioned at the end of Section 2 the order of the operator constructed by the ORTHOGONALSERIES method can sometimes be lowered. We will use the families and the operators introduced in the previous section to do this. Let us consider a linear differential operator L with polynomial coefficients

$$L = \sum_{k=0}^{r} p_k(x) D^k$$

acting on a hypergeometric polynomial series $\sum c_n P_n(x)$, where a family $\{P_n\}$ is determined by the triplet $[\sigma, \tau, B_n]$. Suppose that there exist a HPF $\{P_n^{(i,j)}\}$ (one can always use $\{P_n^{(r,r)}\}$ as we have seen in Section 2) and a recurrence operator $\mathbf{L}$ such that $LP_n = \mathbf{L}P_n^{(i,j)}$. Then by (27)

$$L\left(\sum c_n P_n(x)\right) = \sum c_n \mathbf{L}P_n^{(i,j)}(x) = \sum \mathcal{L}(c_n) P_n^{(i,j)}(x)$$

where $\mathcal{L}$ is the adjoint operator of $\mathbf{L}$. If we are solving $L(\sum c_n P_n) = q$ and $q = \sum_{n=0}^{\infty} \tilde{q}_n P_n^{(i,j)}$, then we obtain the recurrence relation $\mathcal{L}(c_n) = \tilde{q}_n$ for the coefficient sequence. Note that the order of $\mathcal{L}$ is the same as the order of $\mathbf{L}$. If we want to minimize the order of the operator $\mathcal{L}$ we have to choose an appropriate expansion basis $\{P_n^{(i,j)}\}$ and the corresponding recurrence operator $\mathbf{L}$ such that $LP_n = \mathbf{L}P_n^{(i,j)}$ will have the lowest possible order. In the rest of the section we show that the order of $\mathbf{L}$ depends on a chosen basis and is the lowest when the sum $i + j$ is minimal.

Definition 4.1. Let L be a differential operator with polynomial coefficients and $\{P_n\}$ a HPF. A HPF $\{P_n^{(i,j)}\}$ is a *feasible expansion basis* for L if there exists a recurrence operator $\mathbf{L}$ such that $LP_n = \mathbf{L}P_n^{(i,j)}$.
A feasible expansion basis $\{P_n^{(i,j)}\}$ is a *minimal expansion basis* for L if for any other feasible expansion basis $\{P_n^{(k,l)}\}$ for L we have $k + l \geq i + j$.

In view of (36) and (37) it is obvious that if $\{P_n^{(i,j)}\}$ is a feasible expansion basis for L then for any two nonnegative integers k and l the family $\{P_n^{(i+k,j+l)}\}$ is also a feasible expansion basis for L. In the next lemma we show that such a change of the expansion basis will only increase the order of a recurrence operator. Hence, our goal is to reduce the indices of the expansion basis as much as possible. In (29) we established that for a differential operator L of order r $\{P_n^{(r,r)}\}$ is a feasible expansion basis, and we know how to construct operator $\mathbf{L}$ such that $LP_n = \mathbf{L}P_n^{(r,r)}$. But if by chance we have $\mathbf{L} = \tilde{\mathbf{L}}\mathbf{R}_1^{(r-1,r)}$, (36) implies $LP_n = \tilde{\mathbf{L}}P_n^{(r-1,r)}$, so the order of the recurrence reduces by one.

Lemma 4.1. *Let L be a differential operator of order r with polynomial coefficients, let $\{P_n\}$ be a HPF and let $\{P_n^{(i,j)}\}$ be a feasible expansion basis for L such that $i \leq r$ and $j \leq r$. Then $LP_n = \mathbf{L}_1 P_n^{(i,j)}$ and $LP_n = \mathbf{L}P_n^{(r,r)}$ implies $\operatorname{ord}(\mathbf{L}_1) = \operatorname{ord}(\mathbf{L}) - 2r + i + j$.*

Proof. By (36) and (37) we have

$$P_n^{(i,j)} = \mathbf{R}_1^{(i,j)}\mathbf{R}_1^{(i+1,j)}\cdots\mathbf{R}_1^{(r-1,j)}\mathbf{R}_2^{(r,j)}\mathbf{R}_2^{(r,j+1)}\cdots\mathbf{R}_2^{(r,r-1)}P_n^{(r,r)},$$

hence

$$LP_n = \mathbf{L}_1\mathbf{R}_1^{(i,j)}\mathbf{R}_1^{(i+1,j)}\cdots\mathbf{R}_1^{(r-1,j)}\mathbf{R}_2^{(r,j)}\mathbf{R}_2^{(r,j+1)}\cdots\mathbf{R}_2^{(r,r-1)}P_n^{(r,r)}.$$

On the other hand $LP_n = LP_n^{(r,r)}$, therefore we conclude that

$$\mathbf{L} = \mathbf{L}_1\mathbf{R}_1^{(i,j)}\cdots\mathbf{R}_1^{(r-1,j)}\mathbf{R}_2^{(r,j)}\cdots\mathbf{R}_2^{(r,r-1)}.$$

As the operators $\mathbf{R}_k^{(\cdot,\cdot)}$, $k = 1,2$, are all of order one, the result follows. $\square$

Corollary 4.1. *Let* $\{P_n^{(i,j)}\}$ *and* $\{P_n^{(k,l)}\}$ *be feasible expansion bases for* L *and let* $LP_n = \mathbf{L}_1P_n^{(i,j)}$ *and* $LP_n = \mathbf{L}_2P_n^{(k,l)}$. *Then* $\mathrm{ord}(\mathbf{L}_1) < \mathrm{ord}(\mathbf{L}_2)$ *iff* $i+j < k+l$.

This means that the order of a recurrence operator is the lowest when we use the minimal expansion basis for L. Next we show that for a given differential operator only one minimal basis exists.

Lemma 4.2. *Let* $\xi_1 \neq \xi_2$. *If* $\{P_n^{(i,j+1)}\}$ *and* $\{P_n^{(i+1,j)}\}$ *are feasible expansion bases for* L, *then* $\{P_n^{(i,j)}\}$ *is a feasible expansion basis for* L.

Proof. Let $LP_n = \mathbf{L}_1P_n^{(i+1,j)}$ and $LP_n = \mathbf{L}_2P_n^{(i,j+1)}$. Then by (36) and (37) we have $\mathbf{L} := \mathbf{L}_1\mathbf{R}_2^{(i+1,j)} = \mathbf{L}_2\mathbf{R}_1^{(i,j+1)}$. After dividing $\mathbf{L}$ from the right by $\mathbf{R} := \mathbf{R}_1^{(i,j)}\mathbf{R}_2^{(i+1,j)} = \mathbf{R}_2^{(i,j)}\mathbf{R}_1^{(i,j+1)}$ we obtain $\mathbf{L} = \mathbf{KR} + \mathbf{O}$ for some recurrence operators $\mathbf{K}$ and $\mathbf{O}$. As $\mathbf{L}$ and $\mathbf{R}$ are divisible from the right by $\mathbf{R}_1^{(i,j+1)}$ and $\mathbf{R}_2^{(i+1,j)}$ (which are not identical operators), the same is true for $\mathbf{O}$. However, the remainder $\mathbf{O}$ is of order at most one, therefore $\mathbf{O} = 0$, then $\mathbf{L}_1 = \mathbf{KR}_1^{(i,j)}$ and $LP_n = \mathbf{K}P_n^{(i,j)}$. $\square$

Proposition 4.1. *Let* L *be a differential operator with polynomial coefficients and* $\{P_n\}$ *a HPF. Then the minimal expansion basis for* L *is uniquely determined.*

Proof. Let $\{P_n^{(i+k,j)}\}$ and $\{P_n^{(i,j+k)}\}$ with $k \geq 1$ be two minimal expansion bases for L. Then $\{P_n^{(i+k-1,j+k)}\}$ is a feasible expansion basis (FEB) for L. Assume that $\{P_n^{(i+k-1,j+k-m+1)}\}$ where $1 \leq m \leq k$ is a FEB. Since $\{P_n^{(i+k,j)}\}$ is a FEB, so is $\{P_n^{(i+k,j+k-m)}\}$. Hence, by Lemma 4.2, $\{P_n^{(i+k-1,j+k-m)}\}$ is a FEB. By induction on m it follows that $\{P_n^{(i+k-1,j+k-m)}\}$ is a FEB for all $0 \leq m \leq k$. Taking $m = k$ we conclude that $\{P_n^{(i+k-1,j)}\}$ is a FEB, contradicting the minimality of $\{P_n^{(i+k,j)}\}$. $\square$

Therefore the set of all feasible expansion bases for L consists of its minimal expansion basis $\{P_n^{(i,j)}\}$ and all HPF of the form $\{P_n^{(i+k,j+l)}\}$ where k and l are nonnegative integers.

5. Normal form of a differential operator

The next step is to find a form of an operator L that will suggest its minimal expansion basis $\{P_n^{(i,j)}\}$ and the recurrence operator $\mathbf{L}$ such that $LP_n = \mathbf{L}P_n^{(i,j)}$. To this end we will use special differential operators introduced in Section 3 that reduce the expansion basis.

Proposition 5.1. *The following identities hold for* $i, j \in \mathbb{Z}$

$$T_1^{(i-1,j-1)}U^{(i,j)} = U^{(i+1,j-1)}T_1^{(i,j)}$$
$$T_2^{(i-1,j-1)}U^{(i,j)} = U^{(i-1,j+1)}T_2^{(i,j)}$$
$$(x - \xi_1)U^{(i,j)} = U^{(i,j-1)}(x - \xi_1)$$
$$(x - \xi_2)U^{(i,j)} = U^{(i-1,j)}(x - \xi_2)$$
$$(x - \xi_1)T_1^{(i,j)} = T_1^{(i,j-1)}(x - \xi_1)$$
$$(x - \xi_2)T_1^{(i,j)} = T_1^{(i-1,j)}(x - \xi_2) - (x - \xi_1)$$
$$(x - \xi_1)T_2^{(i,j)} = T_2^{(i,j-1)}(x - \xi_1) - (x - \xi_2)$$
$$(x - \xi_2)T_2^{(i,j)} = T_2^{(i-1,j)}(x - \xi_2)$$
$$DU^{(i,j)} = U^{(i+1,j+1)}D + const.$$
$$T_2^{(i+1,j-1)}T_1^{(i,j)} = T_1^{(i-1,j+1)}T_2^{(i,j)} + const.$$
$$DT_k^{(i,j)} = T_k^{(i+1,j+1)}D, \quad for \ k = 1, 2$$
$$(x - \xi_k)D = D(x - \xi_k) - 1, \quad for \ k = 1, 2$$

We omit the straightforward proof.

As an example, look at the first identity of the Proposition 5.1. For both operators $T_1^{(i-1,j-1)}U^{(i,j)}$ and $U^{(i+1,j-1)}T_1^{(i,j)}$, when acting on $P_n^{(i,j)}$ the expansion basis is $\{P_n^{(i,j-2)}\}$. Therefore these "commutation rules" allow us to use operators in generic order. We will denote

$$U^u = U^{(-u-1,-u-1)} \cdots U^{(-1,-1)}U^{(0,0)},$$
$$T_1^{t_1} = T_1^{(-u+t_1-1,-u-t_1+1)} \cdots T_1^{(-u+1,-u-1)}T_1^{(-u,-u)},$$
$$T_2^{t_2} = T_2^{(-u+t_1-t_2+1,-u-t_1+t_2-1)} \cdots T_1^{(-u+t_1-1,-u-t_1+1)}T_1^{(-u+t_1,-u-t_1)},$$

and write

$$Q_r = (x - \xi_2)^{f_2}(x - \xi_1)^{f_1}D^d T_2^{t_2} T_1^{t_1} U^u q_r(x). \tag{40}$$

Differential operator L_r of order r with the leading coefficient

$$p_r(x) = q_r(x)(x - \xi_1)^a (x - \xi_2)^b, \qquad q_r(\xi_k) \neq 0, k = 1, 2, \tag{41}$$

can be written in the form $L_r = Q_r + L_{r-1}$ so that L_{r-1} is of order less than r under the following conditions

$$f_1 + t_1 + u \leq a,$$
$$f_2 + t_2 + u \leq b,$$
$$d + t_1 + t_2 + u = r.$$

The expansion basis for Q_r is $\{P_n^{(i_r,j_r)}\}$ with $i_r = -f_2 + d + t_1 - t_2 - u, j_r = -f_1 + d - t_1 + t_2 - u$. By means of Linear Programming[3] it can be verified that both i_r and j_r are minimal by setting

$$
\begin{aligned}
u &= \min\{a, b, r\}, & d &= r - u - t_1 - t_2, \\
t_1 &= \min\{r - u, a - u\}, & f_1 &= a - u - t_1, \\
t_2 &= \min\{r - u, b - u\}, & f_2 &= b - u - t_2.
\end{aligned}
\tag{42}
$$

In this way among all operators with the leading coefficient (41) we have found the one with the minimal expansion basis.

Definition 5.1. Differential operator L of order r is written in the normal form if $L = \sum_{k=0}^{r} Q_k$, where $Q_k, k = r, r-1, \ldots, 0$, are formed recursively as described in (40) and (42).

Proposition 5.2. If a differential operator $L = \sum_{k=0}^{r} Q_k$ is written in the normal form and $\{P_n^{(i_k,j_k)}\}$ is a minimal expansion basis for Q_k then $\{P_n^{(i,j)}\}$ is the minimal expansion basis for L where

$$i = \max_{0 \leq k \leq r} \{i_k\} \quad and \quad j = \max_{0 \leq k \leq r} \{j_k\}.$$

Proof. Let $\{P_n^{(i,j)}\}$ be the minimal expansion basis for $L_r := L$. By the construction of Q_r and denoting its minimal expansion basis $\{P_n^{(i_r,j_r)}\}$) we have that $i_r \leq i$ and $j_k \leq j$, hence $\{P_n^{(i,j)}\}$ is a feasible expansion basis for $L_{r-1} := L_r - Q_r$. By the same argument $i_{r-1} \leq i, j_{r-1} \leq j, \ldots, i_0 \leq i, j_0 \leq j$ which concludes the proof. $\qquad \square$

Example 5.1. We are looking for a recurrence relation for the coefficients c_n in a Chebyshev series expansion

$$L\left(\sum_{n=0}^{\infty} c_n T_n(x)\right) = 0$$

where $L = (x+1)^2 D^2 - (x+1)D + x + 7/4$. The normal form for L is (we set $\xi_1 = -1, \xi_2 = 1$)

$$L = \left((x+1)D - \frac{3}{2}\right)\left((x+1)D - \frac{1}{2}\right) + x + 1 = T_1^{(1,-1)}T_1^{(0,0)} + x + 1$$

therefore the minimal expansion basis for L is $\{T_n^{(2,-1)}\}$. For the operator $\mathbf{L} = (n - 3/2)(n - 1/2)\mathbf{R}_2^{(2,-2)} + \mathbf{S}_1^{(0,0)}\mathbf{R}_1^{(0,-1)}\mathbf{R}_1^{(1,-1)}$ we find its adjoint operator and after cancelling the denominators we obtain the recurrence of order 3

$$(4n - 2)c_{n+2} + (8n^3 + 28n^2 + 10n - 13)c_{n+1}$$
$$+(2n + 3)(4n^2 - 8n + 1)c_n + 2(2n + 3)c_{n-1} = 0$$

while the recurrence obtained by the Lewanowicz algorithm is of order 4. The algorithm Hyper[18] returns that neither the above recurrence nor its adjoint pair have hypergeometric solutions therefore the obtained recurrence is of minimal order.

5.1. σ has a double root $\xi_1 = \xi_2$

In this case operators $T_1^{(i,j)}$ are not defined (since $\sigma'(\xi_1) = 0$). Therefore we use the operator $T = (x - \xi_1)D$ instead and, by (34) and (38), $TP_n^{(i,j)} = \mathbf{D}^{(i,j)}\mathbf{S}_1^{(i+1,j+1)}P_n^{(i+1,j)}$. The normal form of $L = \sum Q_k$ consists of terms

$$Q_k = (x - \xi_1)^f D^d T_1^t U^u q_k(x)$$

where the leading term of the operator L_k is

$$p_k(x) = q_k(x)(x - \xi_1)^a D^k, \quad q_k(\xi_1) \neq 0$$

and we set

$$u = \min\{k, \lfloor \tfrac{a}{2} \rfloor\}, \qquad d = k - u - t,$$
$$t = \min\{k - u, a - 2u\}, \quad f = a - 2u - t.$$

Example 5.2. Bessel polynomials $Y_m^{(\alpha)}(x)$ are annihilated by (see Ref. 2, pg. 181)

$$L_m = x^2 D^2 + [(\alpha + 2)x + 2]D - m(m + \alpha + 1).$$

As the normal form of L_m with respect to $Y_n^{(\beta)}$ is

$$L_m = DU^{(0,0)} + (\alpha - \beta)T^{(0,0)} - m(m + \alpha + 1),$$

we obtain

$$\mathbf{L}_m = (n + 1)\tau'_{(n-2)/2}\mathbf{R}_1^{(0,0)} + (\alpha - \beta)\mathbf{D}^{(0,0)}\mathbf{S}_1^{(1,1)} - m(m + \alpha + 1)\mathbf{R}_1^{(0,0)}$$

which is of order one. Therefore the connection coefficients $c_{m,n}$ in

$$Y_m^{(\alpha)}(x) = \sum_{n=0}^{m} c_{m,n} Y_n^{(\beta)}(x)$$

satisfy a first order recurrence relation in n.

5.2. σ is of degree one

The associated families are now defined by

$$P_n^{(i)} \longleftarrow [\sigma, \tau + i\sigma', B_n]$$

and we only have operators $\mathbf{R}^{(k)}$ as in (25) to change the basis. When transforming L to the normal form we determine Q_k as

$$Q_k = \sigma^e D^d U^u q_k(x)$$

where we write the leading coeffficient of L_k as $p_k(x) = q_k(x)\sigma^a D^k$ and set

$$u = \min\{a, k\}, \quad d = k - u, \quad e = a - u.$$

Example 5.3. Laguerre polynomials $L_m^{(\alpha)}(x)$ are annihilated by (see Ref. 2, pg. 149)

$$L_m = xD^2 + (\alpha + 1 - x)D + m,$$

and since the normal form of L_m with respect to $L_n^{(\beta)}$ is

$$L_m = DU^{(0,0)} + (\alpha - \beta)D + m + 1,$$

the coefficients $c_{m,n}$ in

$$L_m^{(\alpha)}(x) = \sum_{n=0}^{m} c_{m,n} L_n^{(\beta)}(x)$$

satisfy the first order recurrence relation in n

$$[\alpha - \beta + (1 - E^{-1})(m - n)]c_{m,n} = 0.$$

6. Conclusion

This paper has presented some properties of the hypergeometric polynomials (in the continuous case) that are not exploited in the package ORTHOGONALSERIES and has proven these results in the "hypergeometric way" inspired by the works of A. Nikiforov.

The results will be integrated in a future version of ORTHOGONALSERIES in order to provide more compact results in the generation of recurrences with all the advantages that this represents in applications to connection or linearization problems, or closed form solutions of differential equations.

Acknowledgments

The first author wishes to thank the organizers of the WWCA06 for having invited him to this workshop in honour of Sergei Abramov. He also wants to express his gratefulness to the Symbolic Computation Group (University of Waterloo, Ontario) for its financial support.

References

1. S. A. Abramov and M. van Hoeij. A method for the integration of solutions of ore equations. In W.W. Küchlin, editor, *ISSAC '97*, pages 172–175, New York, 1997. ACM.
2. T. S. Chihara. *An Introduction to Orthogonal Polynomials*. Gordon and Breach, New York, 1978.
3. V. Chvátal. *Linear Programming*. Freeman and Company, New York, 1983.
4. W. Koepf and D. Schmersau. Representations of orthogonal polynomials. *J. Comput. Apll. Math.*, 90:57–94, 1998.
5. C. Lanczos. Trigonomic interpolation of empirical and analytical functions. *J. Math. and Phys.*, 17:123–138, 1938.
6. C. Lanczos. *Applied Analysis*. Prentice-Hall, London, 1956.
7. S. Lewanowicz. Construction of the lowest-order recurrence relation for the jacobi coefficients. *Zastos. Mat.*, 17:655–675, 1983.
8. S. Lewanowicz. Recurrence relations for the coefficients in the jacobi series of linear differential equations. *SIAM J. Math. Anal.*, 17:1037–1052, 1986.
9. S. Lewanowicz. A new approach to the problem of constructing recurrence relations for the jacobi coefficients. *Zastos. Mat.*, 21:303–326, 1991.
10. S. Lewanowicz. Quick construction of recurrence relations for the jacobi co-efficients. *J. Comput. Appl. Math.*, 43:355–372, 1992.
11. S. Lewanowicz. Results on the associated jacobi and gegenbauer polynomials. *J. Comput. Appl. Math.*, 49:137–143, 1993.
12. S. Lewanowicz. Results on the associated classical orthogonal polynomials. *J. Comput. Appl. Math.*, 65:215–231, 1995.
13. S. Lewanowicz. Second-order recurrence relation for the linearization coefficients of the classical orthogonal polynomials. *J. Comput. Appl. Math.*, 69:159–170, 1996.
14. Michael B. Monagan, Keith O. Geddes, K. Michael Heal, George Labahn, Stefan M. Vorkoetter, James McCarron, and Paul DeMarco. *Maple 7 Programming Guide*. Waterloo Maple Inc., Waterloo, Ontario, Canada, 2001. 628 pages.
15. R. J. Ya nez, J. S. Dehesa, and A. F. Nikiforov. The three-term recurrence relation and the differentiation formulas for hypergeometric-type functions. *J. Math. Anal. Appl.*, 188, 1994.
16. A. F. Nikiforov and V. B. Uvarov. *Special'nye Funkcii Matematičeskoj Fiziki*. Nauka, Moskva, 1978.
17. E. L. Ortiz. The tau method. *SIAM J. Numer. Anal.*, 6:480–491, 1969.

18. Marko Petkovsek, Herbert S. Wilf, and Doron Zeilberger. *A=B*. A.K. Peters, Wellesley, Massachusetts, 1996.

19. L. Rebillard. *Etude Théorique et Algorithmique des Séries de Chebyshev Solutions d'équations Différentielles Holonomes*. PhD thesis, Institut National Polytechnique de Grenoble, Grenoble, France, 1998.

20. A. Ronveaux and L. Rebillard. Expansion of multivariable polynomials in products of orthogonal polynomials in one variable. *J. Appl. Math. Comput. (Special issue on Orthogonal Systems and Applications)*, 128:387–414, 2002.

21. A. Ronveaux, A. Zarzo, and E. Godoy. Recurrence relations for connection coefficients between two families of orthogonal polynomials. *J. Comput. Appl. Math.*, 62:67–73, 1995.

22. A. Zarzo, I. Area, E. Godoy, and A. Ronveaux. Results for some inversion problems for classical continuous and discrete orthogonal polynomials. *J. Phys. A: Math. Gen.*, 30, 1997.

ON FACTORIZATION AND SOLUTION OF MULTIDIMENSIONAL LINEAR PARTIAL DIFFERENTIAL EQUATIONS

S.P. TSAREV*

Department of Mathematics
Technische Universität Berlin
Berlin, Germany
E-mail: tsarev@math.tu-berlin.de   sptsarev@mail.ru

We describe a method of obtaining closed-form complete solutions of certain second-order linear partial differential equations with more than two independent variables. This method generalizes the classical method of Laplace transformations of second-order hyperbolic equations in the plane and is based on an idea given by Ulisse Dini in 1902.

Keywords: Integration; Linear partial differential equations.

1. Introduction

Factorization of linear partial differential operators (LPDOs) is often used in modern algorithms for solution of the corresponding differential equations. In the last 10 years a number of new modifications and generalizations of classical algorithms for factorization of LPDOs were given (see e.g. Refs. 2,10,11,13,16,18–20). Such results have also close links with the theory of explicitly integrable nonlinear partial differential equations, cf. Refs. 1,17,22.

As one can see from simple examples (cf. Tsarev[18] and Sec. 2 below) a "naive" definition of factorization of a given LPDO $\hat{L}$ as its representation as a composition $\hat{L} = \hat{L}_1 \circ \hat{L}_2$ of lower-order operators does not enjoy good properties and in general is not related to existence of a complete closed-form solution.

On the other hand, for second-order hyperbolic linear equations in the plane

---

*On leave from: Krasnoyarsk State Pedagogical University, Russia. Partial financial support was provided by RFBR grants 04-01-00130 and 06-01-00814.

we have a well established and deep theory of "generalized" factorization. This theory is known since the end of the XVIII century under the name of Laplace cascade method or Laplace transformations. As proved in Ref. 18, existence of a complete solution of a given second-order hyperbolic equation in the plane in explicit form is equivalent to some "generalized factorizability" of the corresponding operator which in turn is equivalent to finiteness of the chain of Laplace transformations ending in a "naively" factorizable operator. We give a short account of this method in Sec. 2.

There were some attempts to generalize Laplace transformations for higher-order operators or larger number of independent variables, both in the classical time (cf. Refs. 12,14,15) and in the last decade (Refs. 2,19). A general definition of generalized factorization comprising all known practical methods was given in Tsarev.[18] Unfortunately the theoretical considerations of Tsarev[18] did not provide any algorithmic way of establishing generalized factorizability of a given LPDO.

In this paper we move a bit further, extending the algorithmic methods for generalized factorization to the case of second-order operators in the space of more than two independent variables following an approach proposed by Ulisse Dini[4,5] in 1902.

The paper is organized as follows. In the next Section we give an exposition of the classical theory of Laplace transformations. In Sec. 3 we work out an example which demonstrates the idea of Dini transformations and in Sec. 4 we prove a new general result showing that this idea gives a practical method applicable to arbitrary hyperbolic second-order linear equation in the three-dimensional space provided the principal symbol of the operator factors. The last Section is devoted to discussion of algorithmic problems encountered in the theory of Laplace and Dini transformations, their relations to the theoretical basis given by Tsarev.[18] Some conjectures on possibility of generalized factorization and existence of complete solutions in closed form are given.

2. Laplace and generalized Laplace transformations

The cascade method of Laplace (also called the method of Laplace transformations) is until the present date the most general method of finding closed-form complete solutions of hyperbolic second-order linear partial differential equations with two independent variables. Here we briefly sketch this classical theory in a form convenient for our purpose. The complete account may be found for example in Refs. 8,9.

Let a general second-order linear operator L with two independent variables

x and y be given:

$$\hat{L} = \sum_{i=0}^{2} p_i \hat{D}_x^i \hat{D}_y^{2-i} + a_1(x,y)\hat{D}_x + a_2(x,y)\hat{D}_y + c(x,y), \qquad (1)$$

$p_i = p_i(x,y)$, $\hat{D}_x = \partial/\partial x$, $\hat{D}_y = \partial/\partial y$. Hereafter we will always suppose that the operator (1) is strictly hyperbolic, i.e. the characteristic equation $\lambda^2 p_0 - \lambda p_1 + p_2 = 0$ for the principal symbol of (1) has two distinct real roots $\lambda_1(x,y)$, $\lambda_2(x,y)$, so we can introduce two first-order characteristic operators $\hat{X}_i = m_i(x,y)\hat{D}_x + n_i(x,y)\hat{D}_y$, $i = 1,2$, $m_i/n_i = \lambda_i$ ($\hat{X}_i$ are defined up to rescaling $\hat{X}_i \rightarrow \gamma_i(x,y)\hat{X}_i$).
The corresponding equation $Lu = 0$ now may be rewritten in one of two *characteristic forms*:

$$\begin{aligned}(\hat{X}_1\hat{X}_2 + \alpha_1\hat{X}_1 + \alpha_2\hat{X}_2 + \alpha_3)u = \\ (\hat{X}_2\hat{X}_1 + \overline{\alpha}_1\hat{X}_1 + \overline{\alpha}_2\hat{X}_2 + \alpha_3)u = 0,\end{aligned} \qquad (2)$$

where $\alpha_i = \alpha_i(x,y)$. Since the operators $\hat{X}_i$ do not necessarily commute we have to take into consideration in (2) and everywhere below the *commutation law*

$$[\hat{X}_1, \hat{X}_2] = \hat{X}_1\hat{X}_2 - \hat{X}_2\hat{X}_1 = P(x,y)\hat{X}_1 + Q(x,y)\hat{X}_2. \qquad (3)$$

Using the *Laplace invariants* of the operator (2)

$$h = \hat{X}_1(\alpha_1) + \alpha_1\alpha_2 - \alpha_3, \quad k = \hat{X}_2(\overline{\alpha}_2) + \overline{\alpha}_1\overline{\alpha}_2 - \alpha_3,$$

we represent the original operator $\hat{L}$ in two possible partially factorized forms:

$$\hat{L} = (\hat{X}_1 + \alpha_2)(\hat{X}_2 + \alpha_1) - h = (\hat{X}_2 + \overline{\alpha}_1)(\hat{X}_1 + \overline{\alpha}_2) - k. \qquad (4)$$

From these forms we see that the equation $\hat{L}u = 0$ is equivalent to any of the following two first-order systems

$$(S_1): \begin{cases} \hat{X}_2 u = -\alpha_1 u + v, \\ \hat{X}_1 v = hu - \alpha_2 v. \end{cases} \Leftrightarrow (S_2): \begin{cases} \hat{X}_1 u = -\overline{\alpha}_2 u + w, \\ \hat{X}_2 w = ku - \overline{\alpha}_1 w. \end{cases} \qquad (5)$$

If at least one of the Laplace invariants h or k vanishes identically, then the operator $\hat{L}$ factors (in the "naive" way) into composition of two first-order operators and the corresponding system in (5) becomes triangular. So the problem of integration of the original second-order equation is reduced to a much easier problem of integration of linear first-order equations. The latter problem is essentially reducible to finding the complete solution of a (nonlinear!) ODE, see below Section 5.

If $h \neq 0$, $k \neq 0$, one can take one of the systems (5) (to fix the notations we choose the left system (S_1)), express u using the second equation of (S_1)

$$u = (\hat{X}_1 v + \alpha_2 v)/h \qquad (6)$$

and substitute this expression into the first equation of (S_1) in (5); as the result one obtains a X_1-*transformed* equation $\hat{L}_{(1)} v = 0$. It has *different* Laplace invariants (cf. Anderson[1])

$$h_{(1)} = 2h - k\hat{X}_1\hat{X}_2 \ln h + Q\hat{X}_2 \ln h + \hat{X}_2(Q) - \hat{X}_1(P) + 2PQ,$$

$$k_{(1)} = h.$$

If $h_{(1)} = 0$, we can solve this new equation in quadratures and using the same differential substitution (6) we obtain the complete solution of the original equation $\hat{L}u = 0$.

If again $h_{(1)} \neq 0$, apply this X_1-transformation several times, obtaining a sequence of second-order operators $\hat{L}_{(2)}$, $\hat{L}_{(3)}$, ... of the form (2). If on any step we get $h_{(k)} = 0$, we solve the corresponding equation $\hat{L}_{(k)} u_{(k)} = 0$ in quadratures and, using the differential substitutions (6), obtain the complete solution of the original equation. Alternatively one may perform $\hat{X}_2$-*transformations*: rewrite the original equation in the form of the right system (S_2) in (5) and using the substitution $u = (\hat{X}_2 w + \bar{\alpha}_1 w)/k$ obtain the equation $\hat{L}_{(-1)} w = 0$ with Laplace invariants

$$h_{(-1)} = k,$$
$$k_{(-1)} = 2k - h - \hat{X}_2\hat{X}_1 \ln k - P\hat{X}_1 \ln k + \hat{X}_2(Q) - \hat{X}_1(P) + 2PQ. \qquad (7)$$

In fact this $\hat{X}_2$-transformation is a reverse of the $\hat{X}_1$-transformation up to a gauge transformation (see Anderson[1]). So we have (infinite in general) chain of second-order operators

$$\dots \overset{\hat{X}_2}{\leftarrow} \hat{L}_{(-2)} \overset{\hat{X}_2}{\leftarrow} \hat{L}_{(-1)} \overset{\hat{X}_2}{\leftarrow} \hat{L} \overset{\hat{X}_1}{\rightarrow} \hat{L}_{(1)} \overset{\hat{X}_1}{\rightarrow} \hat{L}_{(2)} \overset{\hat{X}_1}{\rightarrow} \dots \qquad (8)$$

As one may prove (see e.g. Goursat[9]) if the chain (8) is finite in both directions (i.e. we have $h_{(N)} = 0$, $h_{(-K)} = 0$ for some $N \geq 0$, $K \geq 0$) one may obtain a quadrature-free expression of the general solution of the original equation:

$$u = c_0 F + c_1 F' + \dots + c_N F^{(N)} + d_0 G + d_1 G' + \dots + d_{K-1} G^{(K-1)} \qquad (9)$$

with definite $c_i(\bar{x}, \bar{y})$, $d_i(\bar{x}, \bar{y})$ and $F(\bar{x})$, $G(\bar{y})$ — two arbitrary functions of the characteristic variables and vice versa: existence of (*a priori* not complete) solution of the form (9) with arbitrary functions F, G of characteristic variables implies $h_{(s)} = 0$, $h_{(-r)} = 0$ for some $s \leq N$, $r \leq K - 1$.

So the complexity of the answer (9) (number of terms in it) is equal to the number of steps necessary to obtain vanishing Laplace invariants in the chain (8) and consequently naively-factorable operators. If (8) is finite in one direction only, one can still obtain a closed-form expression for the complete solution of the original equation; however, it will have *one* of the free functions F or G inside a quadrature expession. More details and complete proofs of these statements may be found in Refs. 8,9 for the case $\hat{X}_1 = \hat{D}_x$, $\hat{X}_2 = \hat{D}_y$, for the general case cf. Goursat [9, p. 30] and Anderson.[1]

Example 1. As a straightforward computation shows, for the equation $u_{xy} - \frac{n(n+1)}{(x+y)^2}u = 0$ the chain (8) is symmetric ($h_{(i)} = h_{(-i-1)}$) and has length n in either direction. So the complexity of the answer (9) may be very high and depends on some arithmetic properties of the coefficients of the operator $\hat{L}$; for the equation $u_{xy} - \frac{c}{(x+y)^2}u = 0$ the chain (8) will be infinite unless the constant $c = n(n+1)$.

Recently a generalization of this classical method was given by Tsarev.[19] In is applicable to strictly hyperbolic linear equations of arbitrary order with two independent variables x, y only.

Dini[4] proposed a simple generalization of Laplace transformations formally applicable to some second-order operators in the space of arbitrary dimension. Namely, suppose that such an operator $\hat{L}$ has its principal symbol

$$Sym = \sum_{i_1, i_2} a_{i_1 i_2}(\vec{x})\hat{D}_{x_{i_1}}\hat{D}_{x_{i_2}}$$

which factors (as a formal polynomial in formal commutative variables $\hat{D}_{x_i}$) into product of two first-order factors: $Sym = \hat{X}_1\hat{X}_2$ (now $\hat{X}_j = \sum_i b_{ij}(\vec{x})\hat{D}_{x_i}$) and moreover the complete operator $\hat{L}$ may be written at least in one of the forms given in (2). This is very restrictive since the two tangent vectors corresponding to the first-order operators $\hat{X}_i$ no longer span the complete tangent space at a generic point ($\vec{x}_0$). (3) is also possible only in the case when these two vectors give an *integrable* two-dimensional distribution of the tangent subplanes in the sense of Frobenius, i.e. when one can make a change of the independent variables ($\vec{x}$) such that $\hat{X}_i$ become parallel to the coordinate plane (x_1, x_2); thus in fact we have an operator $\hat{L}$ with only $\hat{D}_{x_1}$, $\hat{D}_{x_2}$ in it and we have got no really significant generalization of the Laplace method. If one has only (2) but (3) does not hold one can not perform more than one step in the Laplace chain (8) and there is no possibility to get an operator with a zero Laplace invariant (so naively factorizable and solvable).

In the next section we demonstrate, following an approach proposed in another paper by U. Dini,[5] that one can find a better analogue of Laplace transformations for the case when the dimension of the underlying space of independent variables is greater than two. Another particular special transformation was also proposed in Ref. 2,21; it is applicable to systems whose order coincides with the number of independent variables. The results of Ref. 2,21 lie beyond the scope of this paper.

3. Dini transformation: an example

Let us take the following equation:

$$Lu = (\hat{D}_x\hat{D}_y + x\hat{D}_x\hat{D}_z - \hat{D}_z)u = 0. \tag{10}$$

It has three independent derivatives $\hat{D}_x$, $\hat{D}_y$, $\hat{D}_z$, so the Laplace method is *not* applicable. On the other hand its principal symbol splits into product of two first-order factors: $\xi_1\xi_2 + x\xi_1\xi_3 = \xi_1(\xi_2 + x\xi_3)$. This is no longer a typical case for hyperbolic operators in dimension 3; we will use this special feature to introduce two characteristic operators $\hat{X}_1 = \hat{D}_x$, $\hat{X}_2 = \hat{D}_y + x\hat{D}_z$. We have again a nontrivial commutator $[\hat{X}_1, \hat{X}_2] = \hat{D}_z = \hat{X}_3$. The three operators $\hat{X}_i$ span the complete tangent space at every point (x, y, z). Using them one can represent the original second-order operator in one of two partially factorized forms:

$$L = \hat{X}_2\hat{X}_1 - \hat{X}_3 = \hat{X}_1\hat{X}_2 - 2\hat{X}_3.$$

Let us use the first one and transform the equation into a system of two first-order equations:

$$Lu = 0 \Longleftrightarrow \begin{cases} \hat{X}_1 u = v, \\ \hat{X}_3 u = \hat{X}_2 v. \end{cases} \tag{11}$$

Here comes the difference with the classical case $dim = 2$: we can not express u as we did in (6). But we have another obvious possibility instead: cross-differentiating the left hand sides of (11) and using the obvious identity $[\hat{X}_1, \hat{X}_3] = [\hat{D}_x, \hat{D}_z] = 0$ we get $\hat{X}_1\hat{X}_2 v = \hat{D}_x(\hat{D}_y + x\hat{D}_z)v = \hat{X}_3 v = \hat{D}_z v$ or $0 = \hat{D}_x(\hat{D}_y + x\hat{D}_z)v - \hat{D}_z v = (\hat{D}_x\hat{D}_y + x\hat{D}_x\hat{D}_z)v = (\hat{D}_y + x\hat{D}_z)\hat{D}_x v = \hat{X}_2\hat{X}_1 v$.

This is precisely the procedure proposed by Dini.[5] Since it results now in another second-order equation which is "naively" factorizable we easily find its complete solution:

$$v = \int \phi(x, xy - z)\,dx + \psi(y, z)$$

where ϕ and ψ are two arbitrary functions of two variables each; they give the general solutions of the equations $\hat{X}_2\phi = 0$, $\hat{X}_1\psi = 0$.

Now we can find u:

$$u = \int \left(v \, dx + (\hat{D}_y + x\hat{D}_z)v \, dz \right) + \theta(y),$$

where an extra free function θ of one variable appears as a result of integration in (11).

So we have seen that such *Dini transformations* (11) in some cases may produce a complete solution in explicit form for a non-trivial three-dimensional equation (10). This explicit solution can be used to solve initial value problems for (10).

4. Dini transformation: a general result for $dim = 3$, $ord = 2$

Dini did not give any general statement on the range of applicability of his trick. In this section we investigate this question. Obviously one can make different transformations similar to the transformation demonstrated in the previous section, here we concentrate on the simplest case of second-order linear equations with three independent variables *whose principal symbol factors*.

Theorem 4.1. *Let an operator* $L = \sum_{i+j+k \leq 2} a_{ijk}(x,y,z)\hat{D}_x^i \hat{D}_y^j \hat{D}_z^k$ *has a factorizable principal symbol:* $\sum_{i+j+k=2} a_{ijk}(x,y,z)\hat{D}_x^i \hat{D}_y^j \hat{D}_z^k = \hat{S}_1 \hat{S}_2$ *(modulo lower-order operators) with (non-commuting) first-order operators* $\hat{S}_1$, $\hat{S}_2$; $\hat{S}_1 \neq \lambda(x,y,z)\hat{S}_2$. *Then in the generic case there exist two Dini transformations* $L_{(1)}$, $L_{(-1)}$ *of* L.

Proof. One can represent L in two possible ways:

$$L = \hat{S}_1 \hat{S}_2 + \hat{T} + a(x,y,z) = \hat{S}_2 \hat{S}_1 + \hat{U} + a(x,y,z) \tag{12}$$

with some first-order operators $\hat{T}$, $\hat{U}$. We will consider the first one obtaining a transformation of L into an operator $L_{(1)}$ of similar form.

In the generic case the operators $\hat{S}_1$, $\hat{S}_2$, $\hat{T}$ span the complete 3-dimensional tangent space in a neighborhood of a fixed point (x,y,z). Precisely this requirement will be assumed to hold hereafter; operators L with this property will be called *generic*.

Let us fix the coefficients in the expansions of the following commutators:

$$[\hat{S}_2, \hat{T}] = K(x,y,z)\hat{S}_1 + M(x,y,z)\hat{S}_2 + N(x,y,z)\hat{T}. \tag{13}$$

$$[\hat{S}_1, \hat{S}_2] = P(x,y,z)\hat{S}_1 + Q(x,y,z)\hat{S}_2 + R(x,y,z)\hat{T}. \tag{14}$$

First we try to represent the operator in a partially factorized form: $L = (\hat{S}_1 + \alpha)(\hat{S}_2 + \beta) + \hat{V} + b(x, y, z)$ with some indefinite $\alpha = \alpha(x, y, z)$, $\beta = \beta(x, y, z)$ and $\hat{V} = \hat{T} - \beta\hat{S}_1 - \alpha\hat{S}_2$, $b = a - \alpha\beta - \hat{S}_1(\beta)$.

Then introducing $v = (\hat{S}_2 + \beta)u$ we get the corresponding first-order system:

$$Lu = 0 \iff \begin{cases} (\hat{S}_2 + \beta)u = v, \\ (\hat{V} + b)u = -(\hat{S}_1 + \alpha)v. \end{cases} \tag{15}$$

Next we try to eliminate u by cross-differentiating the left hand sides, which gives

$$[(\hat{V} + b), (\hat{S}_2 + \beta)]u = (\hat{S}_2 + \beta)(\hat{S}_1 + \alpha)v + (\hat{V} + b)v. \tag{16}$$

If one wants u to disappear from this new equation one should find out when $[(\hat{V} + b), (\hat{S}_2 + \beta)]u$ can be transformed into an expression involving *only* v, i.e. when this commutator is a linear combination of just two expressions $(\hat{S}_2 + \beta)$ and $(\hat{V} + b)$:

$$[(\hat{V} + b), (\hat{S}_2 + \beta)] = \mu(x, y, z)(\hat{S}_2 + \beta) + \nu(x, y, z)(\hat{V} + b). \tag{17}$$

One can achieve this choosing the free functions $\alpha(x, y, z)$, $\beta(x, y, z)$ appropriately. In fact, expanding the left and right hand sides in (17) in the local basis of the initial fixed operators $\hat{S}_1$, $\hat{S}_2$, $\hat{T}$ and the zeroth-order operator 1 and collecting the coefficients of this expansion, one gets the following system for the unknown functions α, β, μ, ν:

$$\begin{cases} K + \beta P - \hat{S}_2(\beta) = \nu\beta, \\ M - \hat{S}_2(\alpha) + \beta Q = \nu\alpha - \mu, \\ N + \beta R = -\nu, \\ \beta\hat{S}_1(\beta) - \hat{T}(\beta) + \hat{S}_2(a) - \beta\hat{S}_2(\alpha) - \hat{S}_2(\hat{S}_1(\beta)) = -\nu(a - \alpha\beta - \hat{S}_1(\beta)) - \mu\beta. \end{cases}$$

After elimination of ν from its first and third equations we get a first-order non-linear partial differential equation for β:

$$\hat{S}_2(\beta) = \beta^2 R + (N + P)\beta + K. \tag{18}$$

This Riccati-like equation may be transformed into a second-order linear PDE via the standard substitution $\beta = \hat{S}_2(\gamma)/\gamma$. Taking any non-zero solution β of this equation and substituting $\mu = \nu\alpha + \hat{S}_2(\alpha) - \beta Q - M$ (taken from the second equation of the system) into the fourth equation of the system we obtain a first-order linear partial differential equation for α with the first-order term $\beta\hat{S}_2(\alpha)$. Any solution of this equation will give the necessary value of α. Now we can substitute $[(\hat{V} + b), (\hat{S}_2 + \beta)]u = \mu(\hat{S}_2 + \beta)u + \nu(\hat{V} + b)u = \mu v - \nu(\hat{S}_1 + \alpha)v$ into the left hand side of (16) obtaining the transformed equation $L_{(1)}v = 0$.

If we would start the same procedure using the second partial factorization in (12) we would find the other transformed equation $L_{(-1)}w = 0$. □

As a rule neither of the obtained new operators $L_{(1)}$, $L_{(-1)}$ factors into a product of first-order operators as was the case for the operator $L = (\hat{D}_x\hat{D}_y + x\hat{D}_x\hat{D}_z - \hat{D}_z)$ in the previous section. Then one can repeat the described process due to the fact that the principal symbol of the transformed equations is still $\hat{S}_1\hat{S}_2$. Thus we have in the case treated in this section an infinite chain of *Dini transformations*

$$\ldots \leftarrow \ L_{(-2)} \ \leftarrow \ L_{(-1)} \ \leftarrow \ L \ \rightarrow \ L_{(1)} \ \rightarrow \ L_{(2)} \ \rightarrow \ldots$$

If some of the $L_{(i)}$ is factorizable we can obtain its complete solution (under the assumption that one can solve the corresponding first-order equations explicitly) and solving the system (15) w.r.t. u step through this chain back (this again requires solution of linear first-order equations) finally obtaining the complete solution of the original equation $Lu = 0$.

5. Open problems

In the previous three sections we tacitly assumed that the problem of solution of *first-order* linear equations

$$\left(\sum_i b_i(\vec{x})\hat{D}_{x_i} + b_0(\vec{x})\right) u = 0 \tag{19}$$

or

$$\left(\sum_i b_i(\vec{x})\hat{D}_{x_i} + b_0(\vec{x})\right) u = \phi(\vec{x}) \tag{20}$$

can be solved at least for polynomial $b_i(\vec{x})$. In fact it is well known that even in the case $dim = 2$ the problem of complete solution of $(b_1(x,y)\hat{D}_x + b_2(x,y)\hat{D}_y)u = 0$ is equivalent to finding a nontrivial conservation law for the corresponding nonlinear autonomous ODE system or a non-autonomous first-order ODE:

$$\begin{cases} \dfrac{dx}{dt} = b_1(x,y), \\[2mm] \dfrac{dy}{dt} = b_2(x,y), \end{cases} \qquad \Longleftrightarrow \qquad \dfrac{dy}{dx} = \dfrac{b_2(x,y)}{b_1(x,y)} \tag{21}$$

(or, equivalently, finding their general solutions). For polynomial $b_i(x,y)$ this is one of the famous fields of research: study of polynomial vector fields in the plane. Recently an essential advance was made in Refs. 3,6,7; one may hope that a complete algorithm may be found. Still the problem of finding complete solutions of (19) in a suitable "constructive" differential field algorithmically is a challenging problem, as well as the problem of finding solutions for the equation (18).

Another challenging problem is to establish a connection between the general theoretic definition given by Tsarev[18] and the exposed above practical methods based of Laplace and Dini transformations. The known cases suggest the following conjectures presumably valid for operators of arbitrary order and any number of independent variables:

- **Conjecture 1.** If a LPDO is factorizable in the generalized sense of Tsarev,[18] then its principal symbol is factorizable as a commutative polynomial in formal variables $\hat{D}_{x_i}$.

- **Conjecture 2.** If a LPDO of order n is solvable (i.e. the corresponding linear homogeneous equation has an explicit closed-form solution) then its principal symbol splits into product of n linear factors.

One may also suggest to define the principal symbol of a LPDO using different weights for different $\hat{D}_{x_i}$; this would imply for example generalized irreduciblity of parabolic operators similar to $\hat{D}_x^2 - \hat{D}_y$ and potentially provide a powerful criterion of (un)solvability.

One should point out that the *methods* for solution of LPDEs given in the previous sections can not be called completely *algorithmic*: even for the classical case of Laplace transformations and the simplest possible characteristic operators $\hat{X}_1 = \hat{D}_x$, $\hat{X}_2 = \hat{D}_y$ we do not have any bound on the number of steps in the chain (8). Example 1 given in Section 2 suggests that such bounds or other hypothetic stopping criteria would depend on rather fine arithmetic properties of the coefficients of LPDOs.

A more general theoretic treatment suitable for arbitrary (even under- or over-determined systems, cf. Refs. 13,20) based on the language of abelian categories will be exposed in a later publication.

A link to the theory of Darboux-integrable nonlinear PDEs established in Refs. 1,17,22 in our opinion can be extended to other types on nonlinear PDEs. In this connection a generalization of Laplace invariants for higher-dimensional and higher-order cases started in Refs. 2,16,19,21 would be of extreme importance.

Acknowledgments

The author enjoys the occasion to thank the WWCA-2006 organizers for their efforts which guaranteed the success of the Workshop. Certainly the honor of being co-author of a joint paper with S.A. Abramov and our numerous discussions about algorithmic solution of differential equations were (and will always be) highly inspiring!

References

1. I.M. Anderson and N. Kamran, The Variational Bicomplex for Second Order Scalar Partial Differential Equations in the Plane. *Duke Math. J.* **87**, No 2, (1997), 265–319.
2. C. Athorne, A $\mathbf{Z}^2 \times \mathbf{R}^3$ Toda system. *Phys. Lett. A* **206**, (1995), 162–166.
3. Guoting Chen and Yujie Ma, Algorithmic Reduction and Rational Solutions of First Order Algebraic Differential Equations, in *Differential Equations and Symbolic Computation*. Birkhäuser, Trends in Mathematics. (2005), 201–212.
4. U. Dini, Sopra una classe di equazioni a derivate parziali di second'ordine con un numero qualunque di variabili. *Atti Acc. Naz. dei Lincei. Mem. Classe fis., mat., nat.* (ser. 5) **4**, (1901), 121–178. Also *Opere* v. III, 489–566.
5. U. Dini, Sopra una classe di equazioni a derivate parziali di second'ordine. *Atti Acc. Naz. dei Lincei. Mem. Classe fis., mat., nat.* (ser. 5) **4**, (1902), 431–467. Also *Opere* v. III, 613–660.
6. A. Eremenko, Rational solutions of first-order differential equations. *Annales Acad. Sci. Fennicae, Math.* **23**, (1998), 181–190.
7. R. Feng and X.S. Gao, Rational General Solutions of Algebraic Ordinary Differential Equations, in *Proc. ISSAC-2004*, (ACM Press, New York, 2004), 155–162.
8. A.R. Forsyth, *Theory of differential equations*. Part IV, vol. VI. (Cambrudge, 1906).
9. E. Goursat, *Leçons sur l'intégration des équations aux dérivées partielles du seconde ordre a deux variables indépendants.* **2**, (Paris, 1898).
10. D. Grigoriev and F. Schwarz, Factoring and solving linear partial differential equations. *Computing*, **73**, (2004), 179–197.
11. D. Grigoriev and F. Schwarz, Generalized Loewy-Decomposition of D-Modules, in *Proc. ISSAC'2005 (Beijing, China)* ACM Press, (2005), 163–170.
12. J. Le Roux, Extensions de la méthode de Laplace aux équations linéaires aux derivées partielles d'ordre supérieur au second. *Bull. Soc. Math. de France* **27**, (1899), 237–262. A digitized copy is obtainable from http://www.numdam.org/.
13. Z. Li, F. Schwarz and S.P. Tsarev, Factoring systems of linear PDEs with finite-dimensional solution spaces. *J. Symbolic Computation* **36**, (2003), 443–471.
14. Laura Pisati, Sulla estensione del metodo di Laplace alle equazioni differenziali lineari di ordine qualunque con due variabili indipendenti. *Rend. Circ. Matem. Palermo* **20**, (1905), 344–374.
15. L. Petrén, Extension de la méthode de Laplace aux équations

$\sum_{i=0}^{n-1} A_{1i} \frac{\partial^{i+1} z}{\partial x \partial y^i} + \sum_{i=0}^{n} A_{0i} \frac{\partial^i z}{\partial y^i} = 0$. *Lund Univ. Arsskrift* **7**, Nr. 3, (1911), 1–166.

16. E. Shemyakova and F. Winkler, *Obstacles to factorizations of linear partial differential operators.* Submitted, 2006.

17. V.V. Sokolov and A.V. Zhiber, On the Darboux integrable hyperbolic equations. *Phys. Letters A* **208**, (1995), 303–308.

18. S.P. Tsarev, Factorization of linear partial differential operators and Darboux integrability of nonlinear PDEs. *SIGSAM Bulletin* **32**, No. 4, (1998), 21–28; also "Computer Science" e-print cs.SC/9811002 at http://www.arxiv.org/.

19. S.P. Tsarev, Generalized Laplace Transformations and Integration of Hyperbolic Systems of Linear Partial Differential Equations, in *Proc. ISSAC'2005 (July 24–27, 2005, Beijing, China)* ACM Press, (2005), 325–331.

20. Min Wu, *On Solutions of Linear Functional Systems and Factorization of Modules over Laurent-Ore Algebras.* PhD. thesis, Beijing, (2005).

21. H. Yilmaz and C. Athorne, The geometrically invariant form of evolution equations. *J. Phys. A* **35**, (2002), 2619–2625.

22. A.V. Zhiber and V.V. Sokolov, Exactly integrable hyperbolic equations of Liouville type. *Russian Math. Surveys* **56**, No. 1, (2001), 61–101.

TWO FAMILIES OF ALGORITHMS
FOR SYMBOLIC POLYNOMIALS

STEPHEN M. WATT

Ontario Research Centre for Computer Algebra
Department of Computer Science
University of Western Ontario
London Ontario, Canada N6A 5B7
watt@csd.uwo.ca

We consider multivariate polynomials with exponents that are themselves integer-valued multivariate polynomials, and we present algorithms to compute their GCD and factorization. The algorithms fall into two families: algebraic extension methods and interpolation methods. The first family of algorithms uses the algebraic independence of x, x^n, x^{n^2}, x^{nm}, etc, to solve related problems with more indeterminates. Some subtlety is needed to avoid problems with fixed divisors of the exponent polynomials. The second family of algorithms uses evaluation and interpolation of the exponent polynomials. While these methods can run into unlucky evaluation points, in many cases they can be more appealing. Additionally, we also treat the case of symbolic exponents on rational coefficients (e.g. $4^{n^2+n} - 81$) and show how to avoid integer factorization.

1. Introduction

We wish to work with polynomials where the exponents are not known in advance, such as $x^{2n} - 1$. There are various operations we may want to perform, such as squaring the value to get $x^{4n} - 2x^{2n} + 1$, or differentiating it to get $2nx^{2n-1}$. Expressions of this sort arise frequently in practice, for example in the analysis of algorithms, and it is very difficult to work with them effectively in current computer algebra systems.

We may think of these objects as sets of polynomials, one for each value of n, or we may think of them as single values belonging to some new ring. In the ring setting, we wish to perform as many of the usual polynomial operations on these objects as possible. Many computer algebra systems will allow one to work with polynomials with symbolic exponents. They do this, however, either by falling back on some form of weak manipulation of general expressions or by treating all symbolic powers as independent. There are therefore certain operations and simplifications they cannot perform as the relationship between exponents may be non-trivial. We would like, for example, to factorize symbolic polynomials such as

$$x^{n^4-6n^3+11n^2-6(n+2m-3)} - 1000000^m =$$
$$x^{-12m} \times \left(x^{2p} + 10^m x^{p+2m} + 10^{2m} x^{4m}\right) \times \left(x^p + 10^m x^{2m}\right)$$
$$\times \left(x^{2p} - 10^m x^{p+2m} + 10^{2m} x^{4m}\right) \times \left(x^p - 10^m x^{2m}\right)$$
$$p = 1/6\, n^4 - n^3 + 11/6\, n^2 - n + 3$$

and perform operations on symbolic integers

$$16^n - 81^m = (2^n - 3^m)(2^n + 3^m)(2^{2n} + 3^{2m}).$$

This paper examines the problem of working with such symbolic polynomials. The principal contributions are:

- to introduce a useful formulation of symbolic polynomials,
- to show this leads to a well-defined multiplicative structure, with unique factorization
- to present two families of algorithms to compute GCDs, factorizations, *etc.*,
- to extend the notion of symbolic polynomials to allow symbolic operations on the coefficients.

This extends ideas presented in an earlier paper.[7]

The remainder of the paper is organized as follows: Section 2 gives the definition that we shall use as our model for symbolic polynomials. Section 3 discusses the multiplicative properties of symbolic polynomials and shows they have a well-defined unique factorization structure. Section 4 presents a family of algorithms to compute values based on the multiplicative structure of symbolic polynomials. The two examples given are greatest common divisor and factorization. These algorithms are based on the algebraic independence of x, x^n, x^{n^2}, *etc* and work in extensions of polynomial rings. Section 5 presents a second family of algorithms for the same problems, but this time based on projection methods. These methods are based on evaluation and interpolation of the exponent variables. Section 6 addresses technical problems that can arise in term identification in projection methods. Section 7 discusses a number of generalizations of symbolic polynomials. One problem discussed there is treating elements of the coefficient ring with symbolic exponents without having to perform factorizations there. Finally, Section 8 concludes the paper.

2. Symbolic Polynomials

We can imagine a number of models for symbolic polynomials that have desirable properties. Most generally, we could say that any set S, which under an evaluation map gives a polynomial ring $R[x_1, ..., x_v]$, represents symbolic polynomials. This would allow such forms as

$$\gcd(x^n - 1, x^m - 1) + 1 \qquad (1)$$

or

$$(x - 1) \sum_{i=0}^{n} x^i. \qquad (2)$$

Working in terms of explicit ring operations will be useful to us, so we begin by generalizing to symbolic exponents only. This excludes expressions such as (1) and (2).

We recall the concept of a group ring: A *monoid ring* is a ring formed from a ring R and monoid M with elements being the finite formal sums

$$\sum_i r_i m_i, r_i \in R, m_i \in M.$$

A monoid ring has a natural module structure, with basis M, and addition defined in terms of coefficient addition in R. Multiplication is defined to satisfy distributivity, with $r_1 m_1 \times r_2 m_2 = (r_1 r_2)(m_1 m_2)$. When the monoid M is a group, then the algebraic structure is called a *group ring*. For example, the Laurent polynomials with complex coefficients may be constructed as the group ring $\mathbb{C}[\mathbb{Z}]$, viewing $\mathbb{Z}$ as an additive group.

We now define a useful class of symbolic polynomials.

Definition 2.1. The ring of *symbolic polynomials in* $x_1, ..., x_v$ *with exponents in* $n_1, ..., n_p$ *over the coefficient ring* R is the ring consisting of finite sums of the form

$$\sum_i c_i x_1^{e_{i1}} x_2^{e_{i2}} \cdots x_n^{e_{in}}$$

where $c_i \in R$ and $e_{ij} \in \text{Int}_{[n_1, n_2, ..., n_p]}(\mathbb{Z})$. Multiplication is defined by

$$c_1 x_1^{e_{11}} \cdots x_n^{e_{1n}} \quad \times \quad c_2 x_1^{e_{21}} \cdots x_n^{e_{2n}} = c_1 c_2 x_1^{e_{11} + e_{21}} \cdots x_n^{e_{1n} + e_{2n}}$$

We denote this ring $R[n_1, ..., n_p; x_1, ..., x_v]$.

We make use of the integer-valued polynomials, $\text{Int}_{[n_1, ... n_p]}(D)$. For an integral domain D with quotient field K, univariate integer-valued polynomials, usually denoted $\text{Int}(D)$, may be defined as

$$\text{Int}_{[X]}(D) = \{f(X) \mid f(X) \in K[X] \text{ and } f(a) \in D, \text{ for all } a \in D\}$$

For example $\frac{1}{2}n^2 - \frac{1}{2}n \in \text{Int}_{[n]}(\mathbb{Z})$. Integer-valued polynomials have been studied by Ostrowski[3] and Pólya,[5] and we take the obvious multivariate generalization.

Our definition of symbolic polynomials is isomorphic to the group ring $R[(\text{Int}_{[n_1,...,n_p]}(\mathbb{Z}))^v]$. We view $\text{Int}_{[n_1,...n_p]}(\mathbb{Z})$ as an abelian group under addition and use the identification

$$x_1{}^{e_1} x_2{}^{e_2} \cdots x_v{}^{e_v} \cong (e_1, \ldots, e_v) \in (\text{Int}_{[n_1,...,n_w]}(\mathbb{Z}))^v$$

We note that $R[; x_1, ..., x_v] \cong R[x_1, ..., x_v, -x_1, ..., -x_v]$. Also, under any evaluation $\phi : \{n_1, ..., n_p\} \to \mathbb{Z}$, we have

$$\phi : R[n_1, ..., n_p; x_1, ..., x_v] \to R[x_1, ..., x_v, x_1^{-1}, ..., x_v^{-1}].$$

That is, ϕ evaluates symbolic polynomials to Laurent polynomials. It would be possible to construct a model for symbolic polynomials that, under evaluation, had no negative variable exponents. This, however, would require keeping track of cumbersome domain restrictions on the exponent variables. By definition, these symbolic polynomials have a ring structure. What is more interesting is that they also have a useful unique factorization structure that can be computed effectively.

Symbolic polynomials, in the sense we have defined them, can be related to exponential polynomials[2,6] through the transformation $x_i{}^{n^j} \mapsto e^{n^j \log x_i}$. With exponential polynomials, however, it is awkward to capture the notion that the exponents of x_i must be integer valued.

There has also recently been some work on computing Gröbner bases with parametric exponents[4,8] and systems of algebraic equations with parametric exponents.[9] One of the questions asked in this setting is to classify all special cases under evaluation of the parameters. We ask an easier question. Instead, we seek to compute results that are correct under every specialization. This allows us to obtain algorithms for the multiplicative structure of the symbolic polynomials, something that had not been investigated earlier in the parametric setting.

3. Multiplicative Properties

We now show the multiplicative structure of our symbolic polynomials. For simplicity we treat the case when $R = \mathbb{Q}$.

Theorem 3.1. $\mathbb{Q}[n_1, ..., n_p; x_1, ..., x_v]$ *is a UFD, with monomials being units.*

Proof. We first consider the case when exponents are in $\mathbb{Z}[n_1, ..., n_p]$. The fact that $x, x^n, x^{n^2}, ...$ are algebraically independent can be used to remove exponent variables inductively. We observe that

$$x_k^{e_{ik}} = x_k^{\sum_j h_{ij} n_1^j} = \prod_j \left(x_k^{n_1^j}\right)^{h_{ij}} = \prod_j x_{kj}^{h_{ij}}, \quad h_{ij} \in \mathbb{Z}[n_2, ..., n_p].$$

This gives the isomorphism

$$\mathbb{Q}[n_1, n_2, ..., n_p; x_1, ...x_v] \cong$$
$$\mathbb{Q}[n_2, ..., n_p; x_{10}, x_{11}, x_{12}, ...x_{1d_1}, ...x_{v0}, x_{v1}, x_{v2}, ...x_{vd_1}]$$

where d_1 is the maximum degree of n_1 in any exponent polynomial and x_{ij} corresponds to $x_i^{n_1^j}$. Repeating this process p times, we obtain

$$\mathbb{Q}[n_1, n_2, ..., n_p; x_1, ...x_v] \cong \mathbb{Q}[; x_{10...0}, ..., x_{vd_1...d_p}],$$

which is a ring of multivariate Laurent polynomials with the desired properties.

When the exponents come from the integer-valued polynomials $\text{Int}_{[n_1,...,n_p]}(\mathbb{Z})$, as opposed to $\mathbb{Z}[n_1, ...n_p]$, care must be taken to find the fixed divisors of the exponent polynomials. These fixed divisors are given by the content when polynomials are written in a binomial basis. So to show explicitly unique factorization with exponents in $\text{Int}_{[n_1,...,n_p]}(\mathbb{Z})$, we make the change of variables $x_k^{\binom{n_1}{i_1}\cdots\binom{n_p}{i_p}} \to X_{ki_1...i_p}$. Note that the $X_{ki_1...i_p}$ are in one to one correspondence with $x_{ki_1...i_p}$ and so are therefore also algebraically independent. $\square$

Symbolic polynomials can be related to exponential polynomials, which also have a UFD structure.[2]

4. Extension Algorithms

The proof of Theorem 3.1 introduces new variables to replace $x_i^{n_1}$, $x_i^{n_2}$, $x_i^{\binom{n_1}{2}}$, $x_i^{n_1 n_2}$, $x_i^{\binom{n_2}{2}}$, etc. This idea may be used to obtain algorithms for GCD, factorization, square-free decomposition and similar quantities over $\mathbb{Q}[n_1, ...n_p; x_1, ..., x_v]$. We illustrate with algorithms for greatest common divisors and factorization.

Extension Algorithm for Symbolic Polynomial GCD

INPUT: Symbolic polynomials $f_1, f_2 \in \mathbb{Q}[n_1, ...n_p; x_1, ..., x_v]$.
OUTPUT: $g = \gcd(f_1, f_2) \in \mathbb{Q}[n_1, ...n_p; x_1, ..., x_v]$

(1) Put the exponent polynomials of f_1 and f_2 in the basis $\binom{n_i}{j}$.
(2) Construct polynomials $F_1, F_2 \in \mathbb{Q}[X_{10...0}, ..., X_{vd_1...d_p}]$, where d_i is the maximum degree of n_i in any exponent of f_1 or f_2, using the correspondence

$$\gamma : x_k^{\binom{n_1}{i_1}\cdots\binom{n_p}{i_p}} \mapsto X_{ki_1...i_p}.$$

(3) Compute $G = \gcd(F_1, F_2)$.
(4) Compute $g = \gamma^{-1}(G)$.

Under any evaluation map on the exponents, $\phi : \text{Int}_{[n_1,...,n_p]}(\mathbb{Z}) \to \mathbb{Z}$, we have that $\phi(g) \mid \gcd(\phi(f_1), \phi(f_2))$. This g is the maximal uniform gcd in the sense that any other polynomial $g' \in \mathbb{Q}[n_1, ...n_p; x_1, ..., x_v]$ such that $\phi(g') \mid \phi(F_1)$ and $\phi(g') \mid \phi(F_2)$, for all ϕ, also satisfies $g' \mid g$.

Extension Algorithm for Symbolic Polynomial Factorization

INPUT: A symbolic polynomial $f \in \mathbb{Q}[n_1, ...n_p; x_1, ..., x_v]$.
OUTPUT: The factors $g_1, ..., g_n$ such that $\prod_i g_i = f$, unique up to units.

(1) Put the exponent polynomials of f in the basis $\binom{n_i}{j}$.
(2) Construct polynomial $F \in \mathbb{Q}[X_{10...0}, ..., X_{vd_1...d_p}]$, where d_i is the maximum degree of n_i in any exponent of f, using the correspondence

$$\gamma : x_k^{\binom{n_1}{i_1}\cdots\binom{n_p}{i_p}} \mapsto X_{ki_1...i_p}.$$

(3) Compute the factors G_i of F.
(4) Compute $g_i = \gamma^{-1}(G_i)$.

Under any evaluation map on the exponents, $\phi : \text{Int}_{[n_1,...,n_p]}(\mathbb{Z}) \to \mathbb{Z}$, if $\phi(f)$ factors into $f_{\phi 1}, ..., f_{\phi r}$ these factors may be grouped to give the factors $\phi(g_i)$. That is, there is a partition of $\{1, ..., r\}$ into subsets I_i such that $\phi(g_i) = \prod_{j \in I_i} f_{\phi j}$. This factorization into g_i is the maximal uniform factorization in the sense that any other factorization g_i' has $\forall_i \exists_j g_i \mid g_j'$.

It may be that under every evaluation map there is a finer factorization. Erich Kaltofen gives the example $(x^n - 1) \times (y^{n+1} - 1)$. For each n, either the first or second factor is a difference of squares and therefore factors further. There is no further factorization, however, valid for all values of n.

Examples

We use the following pair of polynomials for our examples:

$$p = 8x^{n^2+6n+4+m^2-m} - 2x^{2n^2+7n+2mn}y^{n^2+3n} \tag{3}$$
$$- 3x^{n^2+3n+2mn}y^{n^2+3n} + 12x^{4+m^2-m+2n}$$

$$q = 4x^{n^2+4n+m^2+6m} - 28x^{n^2+8n+m^2+6m+2}y^{4n^2-4n} \tag{4}$$
$$+ 2x^{n^2+4n} - 14x^{n^2+8n+2}y^{4n^2-4n} + 6x^{m^2+6m}$$
$$- 42x^{m^2+6m+4n+2}y^{4n^2-4n} - 21y^{4n^2-4n}x^{4n+2} + 3.$$

We demonstrate the computation of the GCD of p and q and the factorization of p. To begin, we note that the exponents of x in p and q are polynomials in m and n of maximum degree 2. We therefore use

$$\left\{ \binom{n}{i}\binom{m}{j} \;\middle|\; 0 \le i+j \le 2 \right\} = \left\{ 1, n, m, \frac{n(n-1)}{2}, nm, \frac{m(m-1)}{2} \right\}.$$

as a basis for the exponents of x. Likewise we note that the exponents of y are polynomials in n alone and are of maximum degree 2. For them we use the basis

$$\left\{ \binom{n}{i} \;\middle|\; 0 \le i \le 2 \right\} = \left\{ 1, n, \frac{n(n-1)}{2} \right\}.$$

Now we make the change of variables

$$\gamma = \{x \mapsto A, \; x^n \mapsto B, \; x^{\binom{n}{2}} \mapsto C, \; x^m \mapsto D, \; x^{mn} \mapsto E, \; x^{\binom{m}{2}} \mapsto F,$$
$$y \mapsto G, \; y^n \mapsto H, \; y^{\binom{n}{2}} \mapsto I\}$$

to give:

$$p = 8A^4B^7C^2F^2 - 2B^9C^4E^2H^4I^2 - 3B^4C^2E^2H^4I^2 + 12A^4B^2F^2$$

$$q = 4B^5C^2D^7F^2 - 28A^2B^9C^2D^7F^2I^8 + 2B^5C^2 - 14A^2B^9C^2I^8$$
$$+ 6D^7F^2 - 42A^2B^4D^7F^2I^8 - 21A^2B^4I^8 + 3.$$

We then obtain the GCD of p and q as

$$g = 2B^5C^2 + 3$$

and the factorization of p as

$$p = B^2 \times \left(2B^5C^2 + 3\right) \times \left(2A^2F - BCEIH^2\right) \times \left(2A^2F + BCEIH^2\right).$$

Applying γ^{-1}, we have the desired results:

$$g = 2x^{n^2+4n} + 3$$

$$p = x^{2n} \times \left(2x^{n^2+4n} + 3\right)$$

$$\times \left(2x^{1/2\,m^2-1/2\,m+2} - x^{1/2\,n^2+mn+1/2\,n}y^{1/2\,n^2+3/2\,n}\right)$$

$$\times \left(2x^{1/2\,m^2-1/2\,m+2} + x^{1/2\,n^2+mn+1/2\,n}y^{1/2\,n^2+3/2\,n}\right).$$

Remarks

We have described this transformation as though the exponent polynomials were dense, in which case transforming from a power basis to binomial basis introduces no new terms. This is often not the case, so blindly changing to a binomial basis is not always the best strategy.

In the worst case, the number of variables in the new polynomials will be $v(D+1)^p$, where v is the number of base variables, x_i, p is the number of exponent variables, n_i, and D is the degree bound on the n_i in the exponents. In practice, it is often the case that the number of variables occurring in exponents will be small and the exponent polynomials will be of low degree so the introduction of new variables may be acceptable. In other cases, such as when the exponent polynomials are sparse, other approaches may be preferable.

5. Projection Methods

If the number of exponent variables is large and the exponent polynomials are sparse, then it may be advantageous to use an evaluation/interpolation approach. Exponent polynomials may be mapped to integers at several points, the problem solved, and the images combined via interpolation. We illustrate with algorithms for greatest common divisors and factorization.

Projection Algorithm for Symbolic Polynomial GCD (Dense Version)

INPUT: Symbolic polynomials $f_1, f_2 \in \mathbb{Q}[n_1, ...n_p; x_1, ..., x_v]$.
OUTPUT: $g = \gcd(f_1, f_2) \in \mathbb{Q}[n_1, ...n_p; x_1, ..., x_v]$

(1) If $p = 0$ solve problem in $\mathbb{Q}[x_1, ..., x_v, x_1^{-1}, ..., x_v^{-1}]$. Return result.
(2) Let d be the degree bound of n_1 in any exponent of f_1 or f_2.
(3) Choose $d + 1$ distinct evaluation points $e_i \in \mathbb{Z}$.
    Let ϕ_i be the evaluation map $n_1 \mapsto e_i$.
(4) Compute $d + 1$ GCD images $g_i = \gcd(\phi_i(f_1), \phi_i(f_2)) \in \mathbb{Q}[n_2, ..., n_p; x_1, ..., x_v]$ by recursive application of this algorithm.
(5) Identify corresponding terms in the g_i.
(6) Choose one set of corresponding terms and normalize the polynomials so these terms are equal (*e.g.* make those terms 1).
(7) For each set of corresponding terms, interpolate the exponent polynomial to form the corresponding term of g, the GCD.
(8) Return g.

This gives the same GCD as the Extension Algorithm for GCD.

If an evaluation gives a GCD image that is "larger" than the other images, then it is a special case evaluation and should be discarded and another point chosen. If an evaluation point gives a GCD image that is "smaller" than the other images, then the previous evaluations were all unlucky and new points must be chosen.

An important problem is that in step 5 it is not always straightforward to identify corresponding terms. We discuss this in Section 6.

Projection Algorithm for Symbolic Polynomial Factorization (Dense Version)

INPUT: A symbolic polynomial $f \in \mathbb{Q}[n_1, ...n_p; x_1, ..., x_v]$.
OUTPUT: The factors $g_1, ..., g_n$ such that $\prod_i g_i = f$, unique up to units.

(1) If $p = 0$ solve problem in $\mathbb{Q}[x_1, ..., x_v, x_1^{-1}, ..., x_v^{-1}]$. Return result.
(2) Let d be the degree bound of n_1 in any exponent of f.
(3) Choose $d + 1$ distinct evaluation points $e_i \in \mathbb{Z}$.
    Let ϕ_i be the evaluation map $n_1 \mapsto e_i$.
(4) Compute $d + 1$ factorization images $g_{1i} \times \cdots \times g_{ni} = \text{factor}(\phi_i(f)) \in \mathbb{Q}[n_2, ..., n_p; x_1, ..., x_v]$ by recursive application of this algorithm.
(5) Identify corresponding factors in the images, and terms within the factors.

(6) For each set of corresponding polynomial images, choose one set of corresponding terms and normalize the polynomials so these terms are equal.

(7) For each term interpolate the exponent polynomial to form the corresponding term of g_k, the kth factor.

(8) Return $g_1, \cdots, g_n$.

This gives the same factorization, up to units, as the Extension Algorithm for Factorization. As with the GCD computation, there is the problem is that in step 5 it may be difficult to identify corresponding terms. This is discussed later. As with other factorization algorithms, it may be the case that image factorizations have different numbers of factors and that combinations must be tried to form the g_{ki}.

Sparse Algorithms

With naive dense interpolation, a number of problems exponential in the number of variables must be solved in $\mathbb{Q}[x_1, ..., x_v]$. Using sparse interpolation techniques, this is not always necessary. The sparse versions of these algorithms use sparse interpolation of the individual exponent polynomials.

Examples

We use the same p and q as before, defined by equations (3) and (4), and compute the GCD of p and q and the factors of p. The maximum power of m or n in any exponent is 2. For simplicity, we use dense interpolation with $m \in \{1, 2, 3\}$ and $n \in \{1, 2, 3\}$. Letting p_{ij} denote p evaluated at $m = i, n = j$, we have:

$$p_{11} = -2x^{11}y^4 + 8x^{11} - 3x^6y^4 + 12x^6$$
$$p_{12} = -2x^{26}y^{10} + 8x^{20} - 3x^{14}y^{10} + 12x^8$$
$$p_{13} = -2x^{45}y^{18} + 8x^{31} - 3x^{24}y^{18} + 12x^{10}$$
$$p_{21} = -2x^{13}y^4 + 8x^{13} - 3x^8y^4 + 12x^8$$
$$p_{22} = -2x^{30}y^{10} + 8x^{22} - 3x^{18}y^{10} + 12x^{10}$$
$$p_{23} = -2x^{51}y^{18} + 8x^{33} - 3x^{30}y^{18} + 12x^{12}$$
$$p_{31} = -2x^{15}y^4 + 8x^{17} - 3x^{10}y^4 + 12x^{12}$$
$$p_{32} = -2x^{34}y^{10} + 8x^{26} - 3x^{22}y^{10} + 12x^{14}$$
$$p_{33} = -2x^{57}y^{18} + 8x^{37} - 3x^{36}y^{18} + 12x^{16}$$

Similarly, letting q_{ij} denote q evaluated at $m = i, n = j$ gives:

$$q_{11} = 4x^{12} - 28x^{18} + 2x^5 - 14x^{11} + 6x^7 - 42x^{13} + 3 - 21x^6$$

$$q_{12} = 4x^{19} - 28x^{29}y^8 + 2x^{12} - 14x^{22}y^8 + 6x^7 - 42x^{17}y^8 + 3 - 21y^8x^{10}$$

$$q_{13} = 4x^{28} - 28x^{42}y^{24} + 2x^{21} - 14x^{35}y^{24} + 6x^7 - 42x^{21}y^{24} + 3 - 21y^{24}x^{14}$$

$$q_{21} = 4x^{21} - 28x^{27} + 2x^5 - 14x^{11} + 6x^{16} - 42x^{22} + 3 - 21x^6$$

$$q_{22} = 4x^{28} - 28x^{38}y^8 + 2x^{12} - 14x^{22}y^8 + 6x^{16} - 42x^{26}y^8 + 3 - 21y^8x^{10}$$

$$q_{23} = 4x^{37} - 28x^{51}y^{24} + 2x^{21} - 14x^{35}y^{24} + 6x^{16} - 42x^{30}y^{24} + 3 - 21y^{24}x^{14}$$

$$q_{31} = 4x^{32} - 28x^{38} + 2x^5 - 14x^{11} + 6x^{27} - 42x^{33} + 3 - 21x^6$$

$$q_{32} = 4x^{39} - 28x^{49}y^8 + 2x^{12} - 14x^{22}y^8 + 6x^{27} - 42x^{37}y^8 + 3 - 21y^8x^{10}$$

$$q_{33} = 4x^{48} - 28x^{62}y^{24} + 2x^{21} - 14x^{35}y^{24} + 6x^{27} - 42x^{41}y^{24} + 3 - 21y^{24}x^{14}$$

Then we calculate $g_{ij} = \gcd(p_{ij}, q_{ij})$:

$$g_{11} = 2x^5 + 3 \qquad g_{12} = 2x^{12} + 3 \qquad g_{13} = 2x^{21} + 3$$

$$g_{21} = 2x^5 + 3 \qquad g_{22} = 2x^{12} + 3 \qquad g_{23} = 2x^{21} + 3$$

$$g_{31} = 2x^5 + 3 \qquad g_{32} = 2x^{12} + 3 \qquad g_{33} = 2x^{21} + 3$$

This gives one exponent polynomial to interpolate and we obtain

$$g = 2x^{n^2+4n} + 3.$$

We now turn our attention to factoring p. We factor the image polynomials in $\mathbb{Z}[x, y]$:

$$p_{11} = -x^6 \left(y^2 - 2\right) \left(y^2 + 2\right) \left(3 + 2x^5\right)$$

$$p_{12} = -x^8 \left(3 + 2x^{12}\right) \left(x^3y^5 - 2\right) \left(x^3y^5 + 2\right)$$

$$p_{13} = -x^{10} \left(3 + 2x^{21}\right) \left(x^7y^9 - 2\right) \left(x^7y^9 + 2\right)$$

$$p_{21} = -x^8 \left(y^2 - 2\right) \left(y^2 + 2\right) \left(3 + 2x^5\right)$$

$$p_{22} = -x^{10} \left(3 + 2x^{12}\right) \left(x^4y^5 - 2\right) \left(x^4y^5 + 2\right)$$

$$p_{23} = -x^{12} \left(3 + 2x^{21}\right) \left(x^9y^9 - 2\right) \left(x^9y^9 + 2\right)$$

$$p_{31} = x^{10} \left(3 + 2x^5\right) \left(2x - y^2\right) \left(2x + y^2\right)$$

$$p_{32} = -x^{14} \left(3 + 2x^{12}\right) \left(x^4y^5 - 2\right) \left(x^4y^5 + 2\right)$$

$$p_{33} = -x^{16} \left(3 + 2x^{21}\right) \left(x^{10}y^9 - 2\right) \left(x^{10}y^9 + 2\right)$$

We determine which factors correspond by inspection. We let f_1 be the factor with coefficients $\{2, 3\}$, f_2 with $\{\pm1, \mp2\}$, f_3 with $\{1, 2\}$ and u the monomial.

Recall that, in the ring of symbolic polynomials, and in $\mathbb{Q}[x, y, x^{-1}, y^{-1}]$, monomials are invertible and factorization is unique up to units. We pick an arbitrary monomial in each of f_i to be the constant term and normalize. (In principle we could normalize the constant term to 1, but it is convenient here to divide through only by the power product $x^{k_1}y^{k_2}$.) The resulting factors are shown in the following table.

m	n	u	f_1	f_2	f_3
1	1	$-x^6$	$2x^5 + 3$	$y^2 - 2$	$y^2 + 2$
1	2	$-x^8$	$2x^{12} + 3$	$x^3 y^5 - 2$	$x^3 y^5 + 2$
1	3	$-x^{10}$	$2x^{21} + 3$	$x^7 y^9 - 2$	$x^7 y^9 + 2$
2	1	$-x^8$	$2x^5 + 3$	$y^2 - 2$	$y^2 + 2$
2	2	$-x^{10}$	$2x^{12} + 3$	$x^4 y^5 - 2$	$x^4 y^5 + 2$
2	3	$-x^{12}$	$2x^{21} + 3$	$x^9 y^9 - 2$	$x^9 y^9 + 2$
3	1	$-x^{12}$	$2x^5 + 3$	$x^{-1} y^2 - 2$	$x^{-1} y^2 + 2$
3	2	$-x^{14}$	$2x^{12} + 3$	$x^4 y^5 - 2$	$x^4 y^5 + 2$
3	3	$-x^{16}$	$2x^{21} + 3$	$x^{10} y^9 - 2$	$x^{10} y^9 + 2$

Interpolating the exponent polynomials, we obtain

$$u = -x^{4+m^2-m+2n}$$

$$f_1 = 2x^{n^2+4n} + 3$$

$$f_2 = x^{-1/2m^2+mn+1/2n^2+1/2m+1/2n-2}y^{1/2n^2+3/2n} - 2$$

$$f_3 = x^{-1/2m^2+mn+1/2n^2+1/2m+1/2n-2}y^{1/2n^2+3/2n} + 2.$$

This gives the factorization

$$p = u \times f_1 \times f_2 \times f_3,$$

which is the same, up to units, as what we obtained with the extension algorithm. To see this, let $e = m^2 - m + 4$ and multiply u by $-x^{-e}$, f_2 by $-x^{e/2}$ and f_3 by $x^{e/2}$.

6. Finding Corresponding Terms

In general, problems may arise in projection methods when identifying sets of terms for interpolation. In computing GCDs, for example, this amounts to determining which terms correspond in the GCD images. There are three problems that arise:

- The first problem is that, under certain evaluations of the exponent variables, exponent polynomials become equal and terms of the result combine. If there is only one exponent variable, then this can occur for at most $DT(T-1)/2$ evaluation points, where T is the number of terms in the GCD and D is the degree bound on the exponent variable. This is because there are up to $T(T-1)/2$ pairs of distinct exponent polynomials, each pair having at most D common values. For multivariate exponents, terms may combine at an unlimited number of points, but choosing random evaluation points effectively avoids the problem.

- The second problem is that, even if terms do not combine, it may still not be obvious which terms correspond. For example the GCD may have multiple terms with the the the same coefficient and variables. If the coefficient ring is not large enough, then this can occur with high probability.

- The third problem is that one or more evaluation points may give special case results. This is the exceptional case, however. Depending on the problem, the special case results might give an interesting short-cut to a solution or they might be useless and simply be discarded.

In computing factorizations, we have the above problems as well as the usual problem of factor identification.

We illustrate the problem of difficulty identifying corresponding exponents under evaluation with another GCD example, using u and v given as:

$$u = x^{3n^2 - 4n + 8} + 9x^{2n^2 + 4} + x^{n^3 + n^2 - 4n + 4} + 14x^{n^2 + 4n} + 2x^{n^3} \qquad (5)$$

$$v = x^{3n^2 + 8} + 8x^{2n^2 + 4n + 4} + x^{n^3 + n^2 + 4} + 7x^{n^2 + 8n} + x^{n^3 + 4n}. \qquad (6)$$

The exponent polynomials are of degree at most 3, so we evaluate at four points.

$$n = 1 \Rightarrow \qquad \gcd(u, v) = x^6 + 7x^5 + x$$
$$n = 2 \Rightarrow \qquad \gcd(u, v) = 8x^{12} + x^8$$
$$n = 3 \Rightarrow \qquad \gcd(u, v) = x^{27} + x^{22} + 7x^{21}$$
$$n = 4 \Rightarrow \qquad \gcd(u, v) = x^{64} + x^{36} + 7x^{32}.$$

We see that the different evaluations give polynomials with different numbers of terms. It appears that there are three terms in the symbolic polynomial, and that the evaluation at $n = 2$ made two of the exponents equal, giving terms x^{12} and $7x^{12}$.

When the image has three terms, two of the coefficients are the same so it is not clear how to assign the images to symbolic terms for interpolation. Note that the evaluation does not necessarily preserve term order: for $n = 1$ the term with coefficient 7 is of middle degree, for $n = 2$ it is of highest degree and for $n = 3$ and $n = 4$ it is of lowest degree. We must therefore consider the possibility that the terms with coefficient 1 may appear in any order. Thus, even with only two terms having equal coefficients, we have a number of cases to consider exponential in the degree of n. These are shown in the table below. The entries are lists of values for the exponents e_i at $n = [1, 2, 3, 4]$ respectively.

Model	Term 1 $1 \times x^{e_1}$	Term 2 $1 \times x^{e_2}$	Term 3 $7 \times x^{e_3}$
1	[6,12,27,64]	[1,8,22,36]	[5,12,21,32]
2	[6,12,27,36]	[1,8,22,64]	[5,12,21,32]
3	[6,12,22,64]	[1,8,27,36]	[5,12,21,32]
4	[6,12,22,36]	[1,8,27,64]	[5,12,21,32]
5	[6,8,27,64]	[1,12,22,36]	[5,12,21,32]
6	[6,8,27,36]	[1,12,22,64]	[5,12,21,32]
7	[6,8,22,64]	[1,12,27,36]	[5,12,21,32]
8	[6,8,22,36]	[1,12,27,64]	[5,12,21,32]

To discover which is the correct combination, we evaluate at one extra point.

$$n = 5 \Rightarrow \qquad \gcd(u, v) = x^{125} + x^{54} + 7x^{45}.$$

At $n = 5$ either (a) $e_1 = 125$, $e_2 = 54$ or (b) $e_1 = 54$, $e_2 = 125$. Interpolating each model with both choices, we see that model 4 with (b) gives $e_1 = 2n^2 + 4$ and $e_2 = n^3$ with degrees ≤ 3 as required. All other combinations give interpolants of degree 4. We therefore have

$$\gcd(u, v) = x^{2n^2+4} + x^{n^3} + 7x^{n^2+4n}$$

If there are T terms and N evaluation points, then there will be $(T!)^{N-1}$ possible assignments of evaluation points to terms. One of them will give interpolants satisfying the degree bound. Unless T and N are very small, this strategy will obviously be infeasible and another approach will be needed. We also observe that if there is only one exponent variable, then there will be some value beyond which evaluations give images that have a consistent order. This is because a finite set of univariate polynomials will have a finite set of points that make two of the polynomials equal. If this bound can be determined, in principle it avoids the problem of determining which images correspond. In practice, however, it may be too large to be useful (at least in the case of factorization).

Interpolation of Symmetric Functions

There is a better alternative to address the problem of term identification. If there are terms that cannot be distinguished, then we may take advantage of the symmetry and interpolate symmetric functions of the exponent polynomials.

If $t_1, ..., t_T$ are the terms that cannot be distinguished, then we interpolate $S_j(t_1, ..., t_T)$ for different j, where S_j is the j-th elementary symmetric function. We then use *one* evaluation point to break the symmetry and solve for the exponents of the t_i.

We use this method to compute the GCD of u and v given by equations (5) and (6). We wish to determine the exponents of the two terms $x^{A(n)}$ and $x^{B(n)}$, where

$$A(n) = a_3 n^3 + a_2 n^2 + a_1 n + a_0$$
$$B(n) = b_3 n^3 + b_2 n^2 + b_1 n + b_0$$

To do this we interpolate $S_1(A(n), B(n)) = A(n) + B(n)$ and $S_2(A(n), B(n)) = A(n) \times B(n)$. The polynomial for S_2 will be of degree ≤ 6, so we need three extra points. We compute:

$$n = 5 \Rightarrow \qquad \gcd(u, v) = x^{125} + x^{54} + 7x^{45}$$
$$n = 6 \Rightarrow \qquad \gcd(u, v) = x^{216} + x^{76} + 7x^{60}$$
$$n = 7 \Rightarrow \qquad \gcd(u, v) = x^{343} + x^{102} + 7x^{77}$$

We now have

n	$A(n) + B(n)$	$A(n) \times B(n)$
1	$1 + 6$	1×6
2	$8 + 12$	8×12
3	$22 + 27$	22×27
4	$36 + 64$	36×64
5		54×125
6		76×216
7		102×343
Interpolation	$n^3 + 2n^2 + 4$	$2n^5 + 4n^3$

To break the symmetry, we arbitrarily assign $A(1) = 1$ and $B(1) = 6$. Additionally, we equate coefficients in

$$A(n) + B(n) = n^3 + 2n^2 + 4$$
$$A(n) \times B(n) = 2n^5 + 4n^3$$

to obtain 13 equations in the 8 unknowns $\{a_i, b_i\}$. Solving, we obtain:

$$a_0 = 0 \qquad a_1 = 0 \qquad a_2 = 0 \qquad a_3 = 1$$
$$b_0 = 4 \qquad b_1 = 0 \qquad b_2 = 2 \qquad b_3 = 0$$

This determines the two exponents.

7. Generalizations

As mentioned earlier, we may contemplate other algebraic structures to encompass a wider class of expressions. Without going to the most general model of polynomial-valued integer functions, we may consider

- Allowing exponent variables to also appear as regular variables. To do this we can work in $R[n_1, ..., n_p; n_1, ..., n_p, x_1, ..., x_v]$. This is useful if we require formal derivatives.
- Symbolic exponents on coefficients. We discuss these more below.[2,6]
- Non-uniform problems. That is, we may ask how to partition $\mathbb{Z}^p$ as $\bigcup_i D_i$ to obtain more specialized factorizations, gcd, *etc*, valid when restricted to substitutions on individual domains, $\phi : (n_1, ..., n_p) \to D_i$.
- Symbolic polynomials as exponents, or richer structures.
- Other polynomial forms, such as exponential polynomials
- Other problems, *e.g.* Gröbner bases of symbolic polynomials.[8,9]

Let us examine more closely the question of symbolic exponents on coefficients. Suppose we wish to factor a polynomial of the form $x^{4m} - 2^{4n}$. Assuming m and n may take on only integer values, the factorization over $\mathbb{Q}$ is $(x^{2m} + 2^{2n})(x^m + 2^n)(x^m - 2^n)$. This, however is equivalent to $x^{4m} - 16^n$, which is not manifestly the difference of fourth powers. So how can we approach symbolic integer coefficients?

If the coefficient ring is a principal ideal domain, then we may extend our definition to allow symbolic exponents on prime coefficient factors:

Definition 7.1. The ring of *symbolic polynomials in* $x_1, ..., x_v$ *with exponents in* $n_1, n_2, ..., n_p$ *and symbolic coefficients* over the coefficient ring R, a PID with quotient field K, is the ring consisting of finite sums of the form

$$\sum_i k_i \cdot \prod_j c_j^{d_{ij}} \cdot x_1^{e_{i1}} x_2^{e_{i2}} \cdots x_n^{e_{in}}$$

where each product has a finite number of nonzero d_{ij}, $k_i \in K$, c_j are primes $\in R$, $d_{ij} \in \text{Int}_{[n_1, n_2, ..., n_p]}(\mathbb{Z}) \backslash \mathbb{Z}$ and $e_{ij} \in \text{Int}_{[n_1, n_2, ..., n_p]}(\mathbb{Z})$. Multiplication is defined by

$$k_1 c_1^{d_{11}} \cdots c_m^{d_{1m}} x_1^{e_{11}} \cdots x_n^{e_{1n}} \quad \times \quad k_2 c_1^{d_{21}} \cdots c_m^{d_{2m}} x_1^{e_{21}} \cdots x_n^{e_{2n}} =$$
$$k_1 k_2 c_1^{d_{11}+d_{21}} \cdots c_m^{d_{1m}+d_{2m}} x_1^{e_{11}+e_{21}} \cdots x_n^{e_{1n}+e_{2n}}$$

We consider the case of integer coefficients and initially restrict our attention to the situation where the c_j are prime so relationships among symbolic coefficients are apparent. We may use the algebraic independence of p_i^n, $p_i^{n^2}$, *etc* to treat $p_i^{\binom{n_1}{i_1}\cdots\binom{n_k}{i_k}}$ as new variables, as before, in algorithms for factoring, GCD, and related operations.

This straightforward approach requires factoring each integer that appears with a symbolic exponent. In practice we do not want to factor the constant coefficients. Instead, we can form, for any particular problem, a GCD-free basis.[1] For example, if 70^n and 105^n appear, then using the basis $\{X_1 = 2^n, X_2 = 3^n, X_3 = 35^n\}$ avoids factoring. Such a basis may be computed efficiently using only integer GCD and k-th roots.

8. Conclusions

We see a mathematically rich and practically important middle ground between the usual approaches of "symbolic computation" and "computer algebra." In this light, we have explored how to usefully work with symbolic polynomials — polynomial-like objects where the exponents can themselves be integer-valued polynomials.

We have modeled symbolic polynomials using the formal structure of a group ring. These are able to represent the kinds of symbolic polynomials we have seen in practice, for example in the analysis of algorithms. This algebraic structure allows us to perform arithmetic on symbolic polynomials, to simplify and transform them. We find, moreover, a UFD structure that admits algorithms for factorization, GCD, *etc.*

We have sketched two families of algorithms for symbolic polynomials. One puts the exponent polynomials in to a basis that makes their fixed divisors manifest, and then introduces new variables for the symbolic powers. The second family of algorithms is based on evaluation/interpolation, where multiple image problems are solved and the images combined. This approach sometimes has a technical problem in determining which images correspond to do the interpolation. Interpolating symmetric functions of the desired exponent polynomials can avoid some of these difficulties.

We have experimental implementations of both the extension and sparse projection methods, but it is too early to say which method will be most useful in practice.

References

1. E. Bach and J. Shallit, *Algorithmic Number Theory, Volume I: Efficient Algorithms.* MIT Press, 1996.
2. C.W. Henson, L. Rubel and M. Singer, Algebraic Properties of the Ring of General Exponential Polynomials. Complex Variables Theory and Applications, **13**, 1989, 1-20.
3. A. Ostrowski, Über ganzwertige Polynome in algebraischen Zahlköpern, J. Reine Angew. Math., 149 (1919), 117-124.
4. W. Pan and D. Wang. Uniform Gröbner bases for ideals generated by polynomials with parametric exponents. Proc ISSAC 2006, ACM, 269-276.
5. G. Pólya, Über ganzwertige Polynome in algebraischen Zahlköpern, J. Reine Angew. Math., 149 (1919), 97-116.
6. Baron Gaspard Riche de Prony, Essai éxperimental et analytique: sur les lois de la dilatabilité de fluides élastique et sur celles de la force expansive de la vapeur de l'alkool, à différentes températures. Journal de l'École Polytechnique, volume 1, cahier 22, 24-76 (1795).
7. S.M. Watt, Making Computer Algebra More Symbolic. pp 43-49, Proc. Transgressive Computing 2006: A conference in honor of Jean Della Dora, April 24-26, 2006, Granada, Spain.
8. V. Weispfenning, Gröbner bases for binomials with parametric exponents. Technical report, Universität Passau, Germany, 2004.
9. K. Yokoyama, On Systems of Algebraic Equations with Parametric Exponents. pp 312-319, ISSAC '04, July 4-7, 2004, Santander, Spain, ACM Press.

AUTHOR INDEX